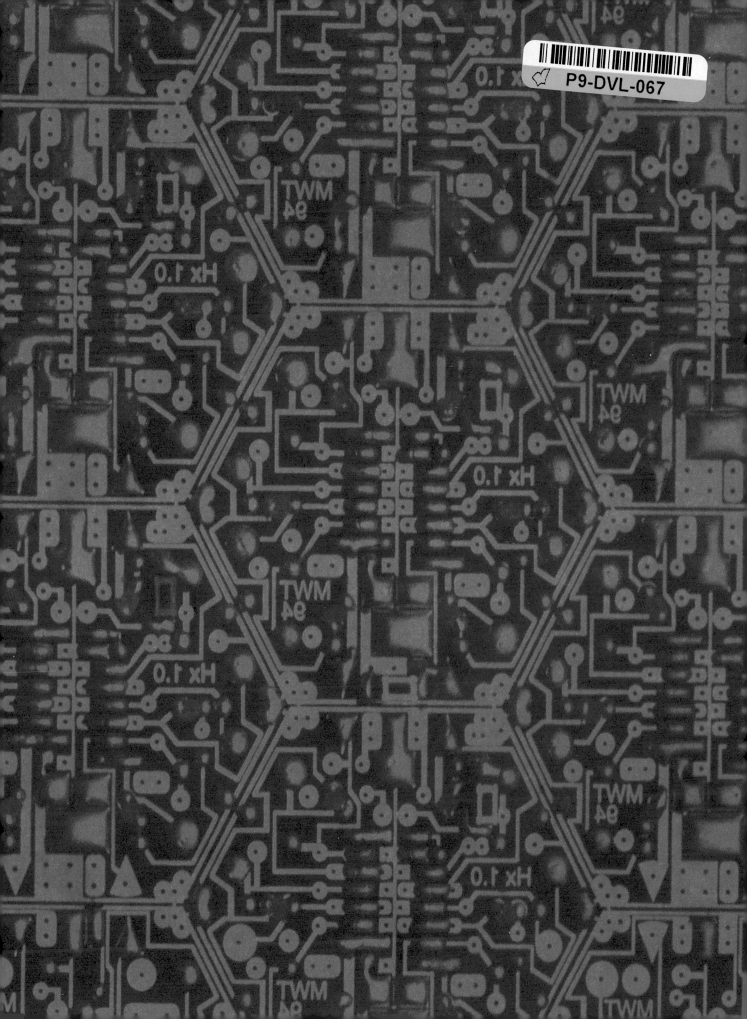

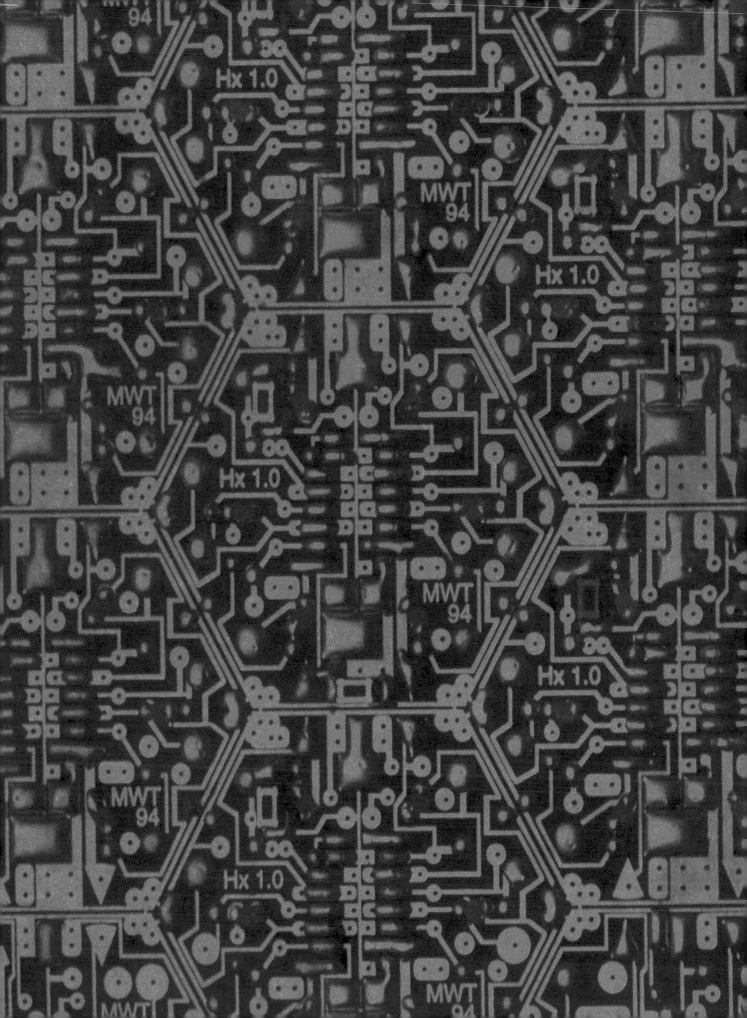

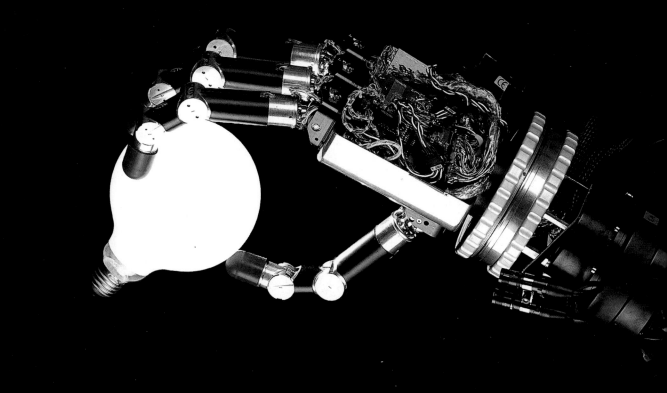

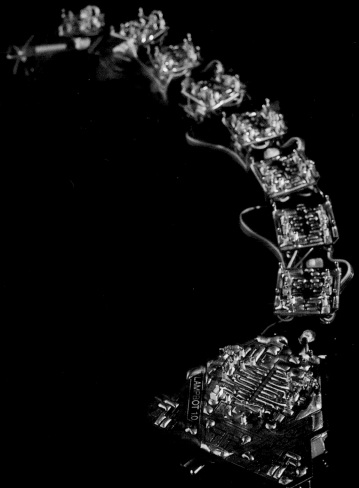

Clockwise from far left: the inner workings of the first-generation face robot from the Hara-Kobayashi Lab in Tokyo; robotic articulated hand from the German Aerospace Center; and Lampbot 1.0, a nine-segment snake robot from researcher Mark Tilden in Los Alamos, New Mexico.

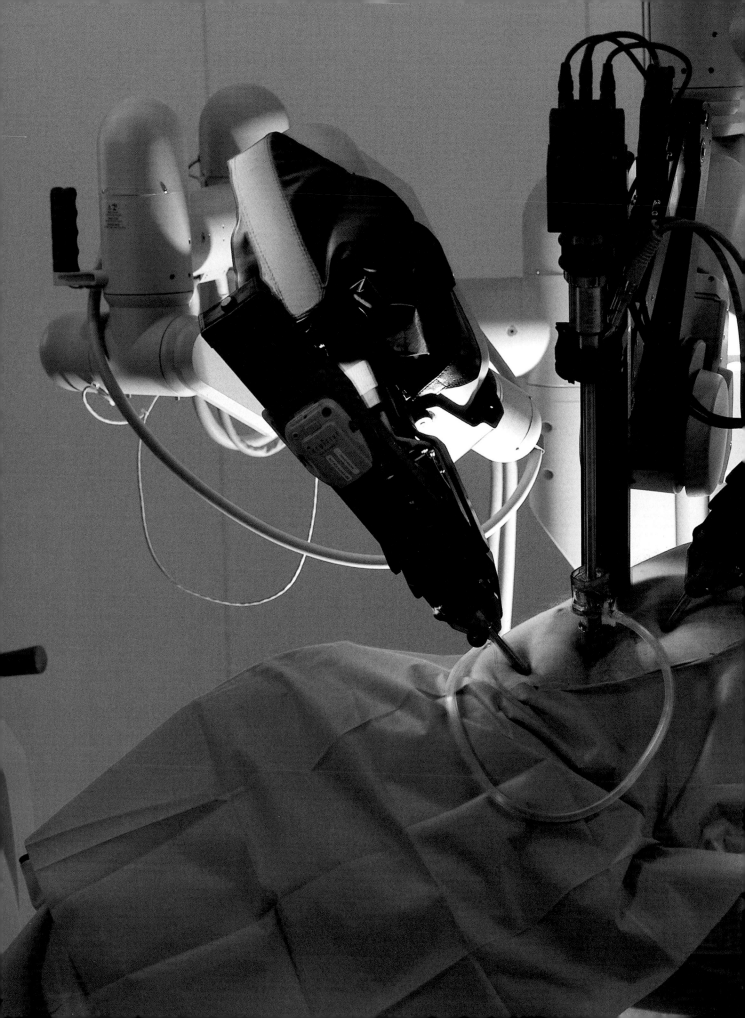

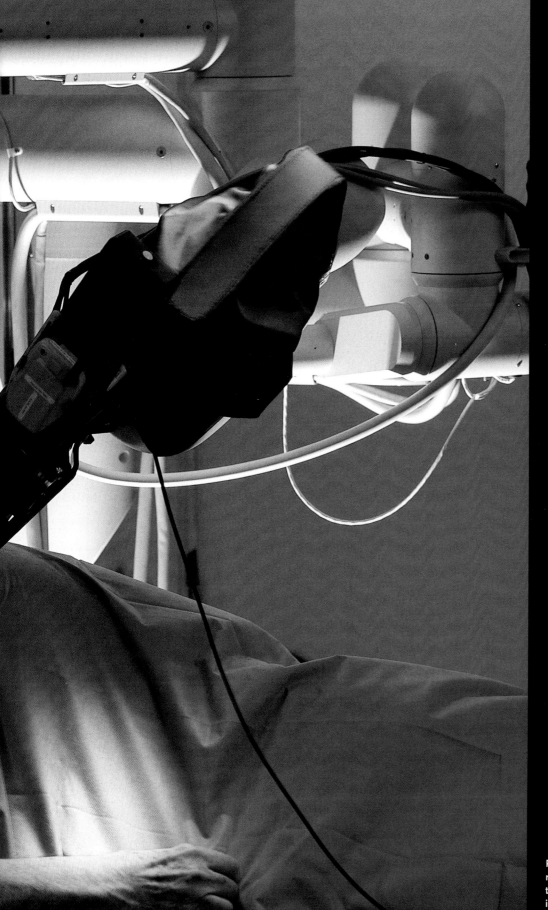

Precision robot arms maneuver
microsurgical instruments
through centimeter-long holes
into the heart of a cadaver in a

Moving like its skittish biological counterpart, Spring Flamingo walks tethered to a boom in a circular course around its home at the MIT Leg Lab.

Robosaurus prowls the parking lot of a Las Vegas casino, showing off its ability to breathe fire and crush cars in its mighty claws. The machine can be rented as a destructive attraction

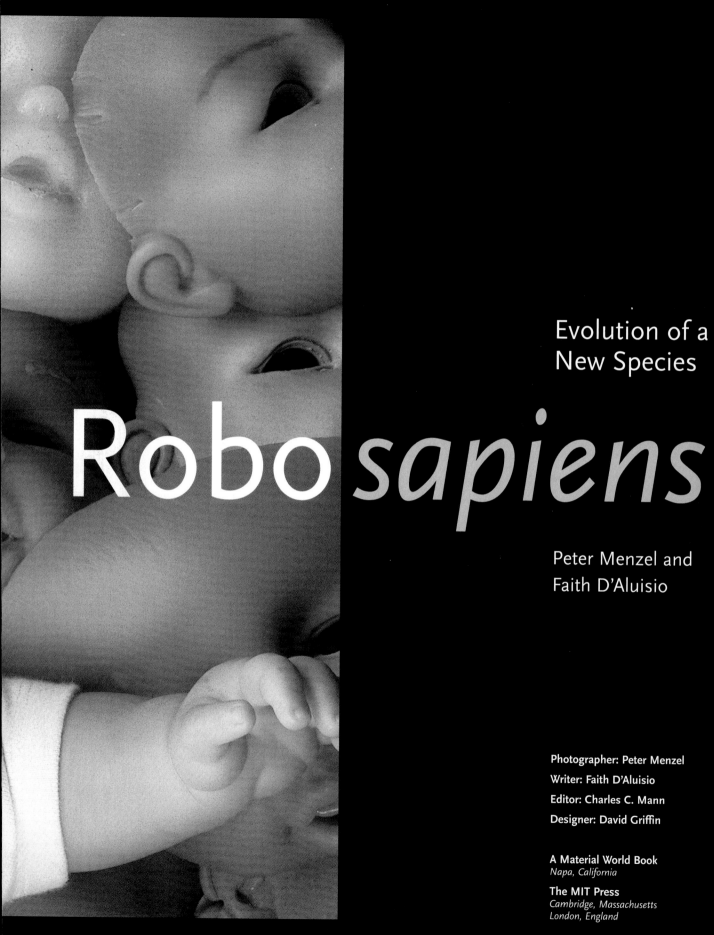

Evolution of a
New Species

Robo *sapiens*

Peter Menzel and
Faith D'Aluisio

Photographer: Peter Menzel

Writer: Faith D'Aluisio

Editor: Charles C. Mann

Designer: David Griffin

A Material World Book
Napa, California

The MIT Press
Cambridge, Massachusetts
London, England

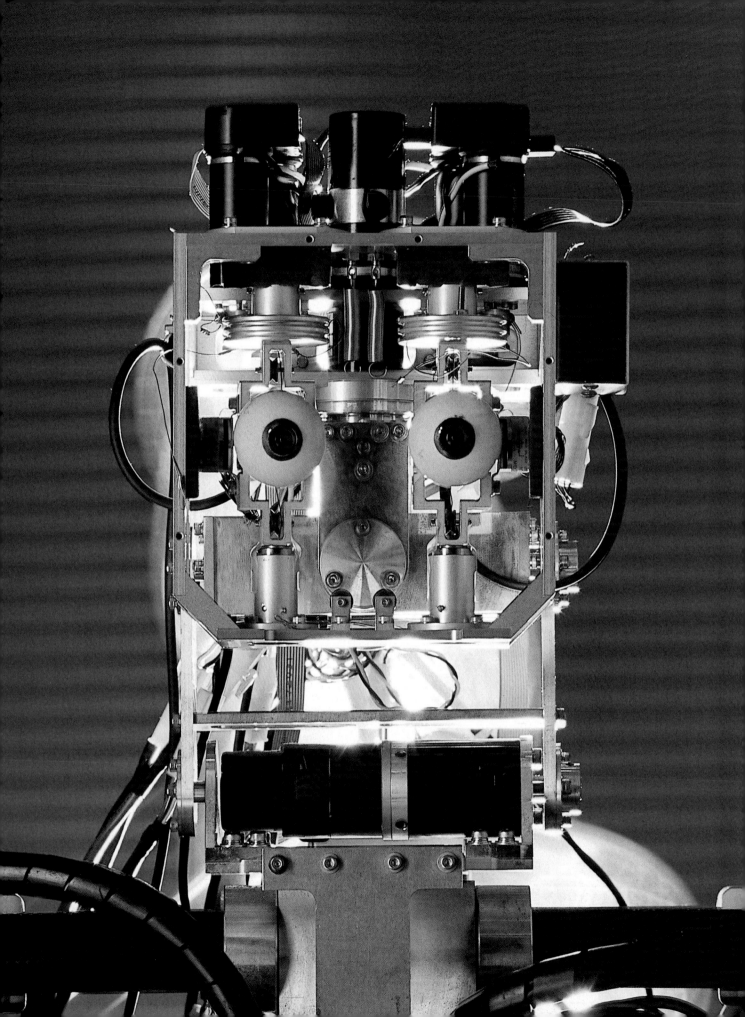

Contents

Introduction Peter Menzel 16

Electric *dreams* 20

Robo *sapiens* 34

Bio *logical* 84

Remote *possibilities* 122

Work *mates* 162

Serious *fun* 198

Methodology Faith D'Aluisio 232

Glossary 234
Recommended Reading 236
Index 238
Acknowledgments 239

In a years-long quest, students at Waseda University in Tokyo are constantly tweaking the programming of WABIAN R-II in the hope of making the heavy, two-meter-tall machine walk as easily as a human being.

PREVIOUS PAGES

BIT—Baby It—is the prototype for My Real Baby, the most sophisticated robot doll yet made. According to a press release, it is only the "first born" in a series of dolls created from the union of its parent companies, toy giant Hasbro and iRobot, a small Massachusetts robotics firm.

Introduction *Peter Menzel*

Robot: n (Czech, from *robota*, compulsory labor) 1. A machine that looks like a human being and performs various complex acts (as walking and talking) of a human being. 2. A mechanism guided by automatic controls. [Merriam-Webster Collegiate Dictionary, 1999]

Robo sapiens: n (English, from *robot*, a mechanism guided by automatic controls; and Latin, from *Homo sapiens*, mankind) 1. A hybrid species of human and robot with intelligence vastly superior to that of purely biological mankind; began to emerge in the twenty-first century. 2. The dominant species in the solar system of Earth. [Microsoft Universal Dictionary, 2099]

Before Faith and I began this book I would have attributed the term *Robo sapiens* to a science-fiction writer. I would have been amused, but would have scoffed especially hard at those modifiers "vastly superior" and "dominant" in its (admittedly hypothetical) definition. I have been skeptical about technology's undelivered promises ever since I began a career as a photojournalist, more than two decades ago. More specifically, I am skeptical not about technology per se but about the way societies misuse and misunderstand it. Nuclear power, for example, is a wonderful technology that would have had a prominent place in any wise, far-seeing, and incorruptible society. In the future, I hope to encounter such a society.

Pessimism about society's potential to misuse technology is nothing new. Czech writer Karel Čapek coined the term "robot" in a play he wrote in 1920 called *R.U.R.* [first performed in 1923]. The play's dark plot revolves around a factory—Rossum's Universal Robots, the *R.U.R.* of the title—that populates the world with artificial slaves, meant to relieve humans of the drudgery of work. Built in ever-increasing numbers and with expanding intelligence, they soon outnumber their human masters, and then they are used as soldiers. Eventually, a robot revolt wipes out the human race. It's interesting that the person who invented the modern concept of robots predicted that they would destroy us all.

A year ago, when our friend Thomas Borchert, technology and science editor at *Stern* magazine in Germany, asked me to document advances in robotics and artificial intelligence around the world, I was eager to seek out the planet's most advanced machines and their makers (he didn't mention any threat to the human race). My interest—and my skepticism—increased when I learned that roboticists were predicting that human intelligence would soon be surpassed by machine intelligence—if not in the next decade, then in the decade after.

How much was hype? Was there really a revolution in robotics, as the roboticists claimed? I knew that robotics had not lived up to the dreams of its pioneers in the 1960s and 1970s. Today, years after robots were supposed to be on the streets and in our homes, the machines are largely unseen. The only robots widely in use are computer-driven slaves bolted to factory floors and laboratory benches. Knowing this, I did not expect to be intellectually kidnapped by a robot. But then it happened, in Japan.

We were shooting a private demo of the Honda P3—a robot that looks like an astronaut in a white spacesuit—at the car company's top-secret development center outside Tokyo. (Yes, the Japanese automaker had quietly spent millions to build a bipedal robot.) Watched proudly by a host of attendant technicians, the robot walked down a path on its own two feet, opened a door, stepped through it, closed the door behind itself, then gracefully climbed a flight of stairs. Was there a little man inside? No. Just electric motors and electronics. We knew that the Honda humanoid was not autonomous—that every step was programmed—but we were still amazed that a machine could maneuver like a human.

At that time, we were nearly halfway through a month-long assignment. We had seen a number of very advanced robots, but most worked haltingly, and some only after interminable delays to load software or troubleshoot a fragile connection. No single machine had yet captured our imagination enough to make us believe the glowing predictions of robotic evolution—in fact, many machines were quite primitive beasts. But when we saw the Honda humanoid's jaunty demo, we realized that with enough time and money, layers of complexity could be added to any of the other nonhuman-looking robots as well—upgrading them from clumsy class into the realm of smooth operators.

The Honda robot is not alone. As we learned more and saw more, we realized that twenty years after their initial predictions, the futurists might at last be proven right. Machine intelligence is here, in infant form. And now it's learning to walk and talk.

After completing an additional ten-day robot shoot in Europe in the spring, we decided to continue the grand tour of robotics labs around the world, visiting and revisiting labs and talking to the people who want to build our mechanical future. We wanted to know where the robots were going. More than that, we wanted to know where we—that is, humankind—were going. We were amazed by what we found.

We discovered that the dawn of our postbiological future may arrive sooner than we imagine. Consider technological progress to date. It took billions of years for primitive life forms to evolve into mammals, millions of years for the descendents of those first mammals to lumber into the Stone Age, thousands of years for humans to advance from stone to steel—and only sixty-six years for people to soar from Kitty Hawk to a stroll on the moon. We accelerated from the Wright stuff to the right stuff so quickly that the question inevitably arises: how long—or how short—a time will it be to the next step, *Robo sapiens*?

One part of the answer is that we are already part way there—indeed, we began evolving in that direction in the last century. With our artificial hips, prosthetic knees, false teeth, hearing aids, pacemakers, breast and penile implants, we are part cyborg today. (Well, not me personally, you understand.) That list of mechanical parts doesn't include transplanted organs, skin grafts, and plastic surgery. Or the new mechanical hearts that are under development today. Or the cochlear and retinal implants that will be here tomorrow. There may be general resistance to implanting chips in people's brains, but when a bio-chip is developed to easily enhance memory or linguistic skills or mathematical abilities—how long will people just say no? Kevin Warwick (see page 29) plans to address this question by implanting a chip in his own body next year. It could be true: the next step in human evolution could indeed be from man to machine.

Another part of this puzzle is provided by the silicon chip, which has rammed our technological future into overdrive. Computers, nearly ubiquitous now, soon will be. Every year, they get smaller, faster, cheaper, more powerful. According to what is known as Moore's Law, computer chips will become faster and more powerful by a factor of two every year or so. This exponential increase shows no signs of abating, which means that chips will get faster and faster even as they get smaller and smaller. Contrast this with the evolution of the human brain and many experts conclude that machine intelligence will inevitably surpass human intelligence—the only question, is what will happen when it does.

These are the unknowns in the equation that make the future exciting, and possibly a bit scary—they blur the lines between science and philosophy. Since we don't understand the basis for human consciousness yet, how can we create it in machines? Or will mechanical minds start up by themselves when they reach a certain level of complexity? (Is there critical thought, like critical mass?) And if machine intelligence does jump-start itself, will we be able to turn it off if we don't like the results? If an artificial intelligence is composed of human-compiled facts, does this mean it will be kindly disposed to

Snarling at the rush-hour traffic, this new animatronic—that is, lifelike and electronic—replica of an *Allosaurus* is returning from the paint shop to the Dinamation factory in Orange County, California.

A technophobic view of the future places humans on the same highway to extinction—driven there not by a cataclysmic meteor crash, but by the impact of robots with intelligence vastly superior to that of purely biological mankind.

humans? Or will it be ethnocentric (or should I say mechanocentric)? Could it threaten us?

But this assumes that all the evolution will be on the mechanical side. What if we could start from early childhood with the complete storehouse of knowledge from our predecessors rather than having to reinstall it each time in our organic-hard-drive brains? Herein lies the lure of robotic silicon intelligence—bio-chips (brain chips, one might call them). Electronic memory can be accessed a million times faster than human synapses—it does not have to sleep, and can be downloaded to others (similar machines, or people with compatible chips). It can be stored, compressed, sorted—even spindled and mutilated. (No need to get angry, greedy, or jealous.) Of course this downloaded information would only be data, and data is not knowledge, let alone wisdom. But in a postbiological world, how long would it be before we learned how to take the next step and download wisdom? (Could it be a program?)

Now factor in the realm of the very small, where nanotech and bio-chips promise things unheard of (in fact, undreamed of). Micro Air Vehicles, on page 156, are already here—robotic flies are on the way. Kris Pister's smart dust (see page 26) may seem like science fiction, but for how long? Sir Arthur C. Clarke wrote, "Any sufficiently advanced technology is indistinguishable from magic." Today, the technological magic is more than accepted, it is expected. We believe that our dreams are not just dreams, they are sneak previews.

Atsuo Takanishi of the Humanoid Research Laboratory, Waseda University, Tokyo, conversing with Faith D'Aluisio at his university laboratory.

In one such preview of primordial mechano-motion, we watched Ariel, an eight-legged crab robot, sidle along the shore of a wooded pond in Massachusetts (see page 100). Probably the most complex self-contained machine capable of underwater ambling, Ariel is still very limited in what it can do. When I mentioned this to Ed Williams, the robot wrangler technician who was demonstrating the machine, he Wrightly countered, "What could an airplane do in 1920? Get you killed in it, that's about all."

The most exciting part of our journey was bearing witness to the first halting steps— spasmodic evolutionary spurts—of robotics at the beginning of the new millennium. To be sure, there were glitches along the way, and there will be more. At the MIT Leg Lab, we saw a robotic *Troodon* dinosaur (see page 114) twitch its way up to a standing position, ready to walk forth, only to be crippled by a software glitch before it took a single step. Mimicking the incredibly complex biological systems that took millions of years to evolve—and to survive and function in a narrow niche—is a daunting challenge. Shortcutting this process with high-tech tools to analyze animal locomotion, Robert J. Full's biology lab at UC Berkeley (see page 90) provides data that roboticists at Stanford, Michigan, and McGill use to build robots that translate these movements into mechanical systems guided by the genius of biological evolution.

The ultimate quest, the Grail of many roboticists today, is to build a humanoid robot. The Honda P3 that so astounded me is but one of many attempts to reach this goal. The quest is moving along at a steady pace that will accelerate with advances in materials, computing power, and electromechanical interfaces. But why build humanoids? Walking upright on two legs is as difficult as flying. Is there any other living creature more unstable than a human in an ambulatory position? Many roboticists say the reason to build such an unwieldy being is psychological—they believe that robots will be more easily accepted by humans if they are built in the humans' own image. Others think it folly; they admit the motive for making a machine in one's own image may be psychological, but see it more as a fascination with playing God. The well-known MIT roboticist Rod Brooks (see page 58) frankly explained that he jumped from building insect robots to Cog, a humanoid, because he didn't want to spend a lifetime crawling along the robotic evolutionary path when he sensed he might be able to have a run at something approaching an android.

Shigeo Hirose, one of Japan's most respected roboticists, thinks the humanoid shape may not be the best idea, in engineering terms, but he argues that any robot engineered to be intelligent could be engineered to be moral. Robots could be saints, he told us. We could build them to be unselfish, because they don't have to fight for their biological existence (see page 89). They can download their artificial intelligence to another machine, thereby continuing their "life." They don't have to be like the rampaging machines in *R.U.R.* Nice robot, smart robot, saintly robot.

In spirit, it seems we are more ready for robots than they are for us. Despite our fears of Frankenstein's monster or Hollywood's Terminator, if something robotic can make our life better, we embrace it. We already accept and expect robotic systems to help us: power steering, cruise control, and GPS systems in our cars; autopilot and IFR landing systems in airplanes. We trust these machines with our lives—we have built them with multiple levels of checks and safety features. Our comfort level with robots is rising, too. Many swimming-pool owners have them cleaning the bot-

Photo: Alexander J. Wright

tom of the pool every day, and several companies are developing robot vacuum cleaners for the home. Robots already work in many factories—there is no resistance, really. In the General Motors plant that we visited (see page 190), robots do all the dirty work, and workers welcome the relief. They say they don't fear job loss from robots—if a robot takes over their position on the factory floor, they get to be transferred to less physically demanding work within a more productive factory.

Robotic progress today is poised to take off like the personal computer did a decade ago. The shortcomings that have kept robot research away from millions of small inventors and researchers—no universally accepted operating control system and a lack of standardized parts—should soon be complaints of the past.

After cutting off half the face of BIT, a prototype robotic doll, photographer Peter Menzel is himself photographed at the headquarters of the toy's designer, iRobot of Somerville, Massachusetts.

A standard operating control system and new lightweight, compliant materials (see page 98) on shape-deposition manufacturing), coupled with more powerful, energy-efficient actuators, could bring robot design into the wide, swift mainstream of everyday science.

Today's robots are more than factory workers; they are explorers, space laborers, surgeons, maids, actors, pets—the list gets longer every day. We should expect to be surprised, because our imagination will create many, many more roles. Our mechanical destiny is not to be denied, and the questions arising from the creation of these creatures are ones that will shape the future of humanity, in whatever form it eventually assumes.

This book is not meant to be Genesis-like, detailing each mechanical iteration. Instead, I hope that my camera and Faith's tape recorder have produced a field guide to our mechanical future that will make the passage less frightening. Should we assume the worst: that *Robo sapiens* will eventually run amok like the machines in *R.U.R.*? In Act III, during the robots' siege of the factory, the optimistically naïve Helena, one of the last surviving humans, tragically laments to Radius, a robot leader, that his intelligence should engender understanding, not conflict. In the 1920s play, it didn't. In tomorrow's real world, we hope it will. Radius can be a saint—if we understand the process that created him and the things in ourselves that drove us to do it.

HELENA. Doctor Gall gave you a larger brain than the rest, larger than ours, the largest in the world. You are not like the other Robots, Radius. You understand me perfectly.
RADIUS. I don't want any master. I know everything for myself.　　　　—R.U.R. *by Karel Čapek, 1920*

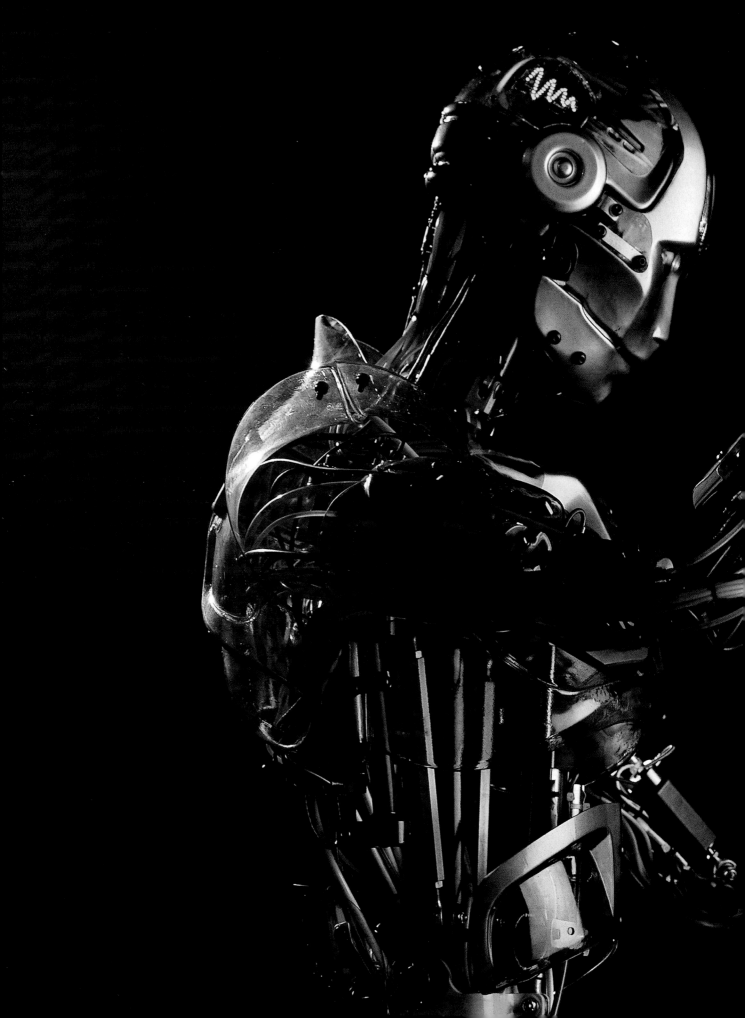

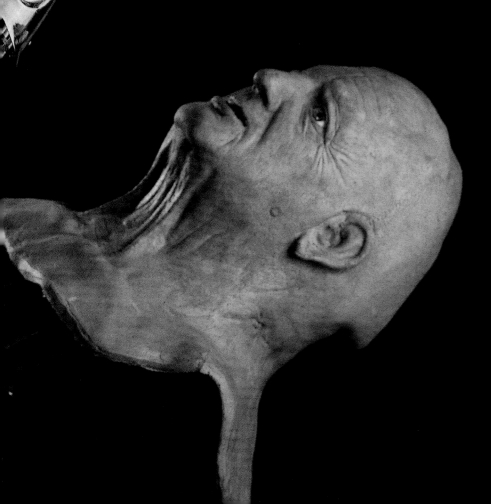

Electric *dreams*

Wielding a paint brush, a robot touches up its human master in this photo-illustration at the SARCOS robot company in Salt Lake City. Some roboticists believe that machines will never approach human abilities; others, that they will inevitably take over the world. Still a third school argues that these scientists have it all wrong. Robots will neither fall short of people nor overwhelm them. Instead, people will become robots, electronically merging the extraordinary consciousness of *Homo sapiens* and the almost infinitely durable bodies of robots: *Robo sapiens*.

Electric Dreams: What the future may hold

There are great and wondrous robots in our future, say *Those Who Know*. Robots will assist the elderly and infirm into and out of wheelchairs and beds, be conversant in several languages, intuit despair, watch over babies, and provide a sympathetic ear to the lonely. Smartly appointed robotic vacuum cleaners, robotic cars, robotic maids, robotic cleanup squads, and robotic personal assistants will lead to greater efficiency and safety in the world, working where humans can't or won't and providing more free time for their masters. All but human, ubiquitous, they will be woven invisibly into the fabric of our lives.

There are terrible times in store for the human race, say *Those Who Know*. Robots will begin as elder-care assistants and vacuum cleaners, but they won't stay there. They will take what we teach them and learn to want more. They will make themselves ever smarter and stronger, until finally, discovering that they are better than we are at everything we do, they will refuse to take our orders. Far from becoming indispensable components of our lives, they will find our lives ever more unnecessary to them. If they don't end up ignoring us, they'll eliminate us.

Who are *Those Who Know?* The robot pundits. The prognosticators of our mechanical future. The digital soothsayers. Academic and corporate researchers, for the most part, they study such exotic domains as artificial intelligence, cyberneurology, and biomimetics. Often at odds with one another but never unsure of their auguries, they claim to know the future of the human race, and to know that it will involve robots. Robots, they all agree, will transform the future. The problem is; they differ on the details. Like whether robots will serve us—or we will serve them.

For more than a year, Peter Menzel and I explored robotics laboratories in Europe, Asia, and America, looking at projects in development and speaking with researchers. Over time, we concluded that these pundits are at least partly right. Clearly, the robots are coming. Although the machines we saw were often barely functional, they were gaining in capacity. The discipline of robotics—a quirky union involving the fields of artificial intelligence, computer science, mechanical engineering, psychology, anatomy, and half a dozen others—is perhaps moving faster than even the researchers know.

The discipline is advancing so rapidly, in fact, that some roboticists have begun questioning the direction in which their work is heading. Not every roboticist we encountered felt inclined to speculate on the future, of course. Like all branches of science and engineering, robotics is full of researchers who try to focus entirely on their work in the present. For the most part, these people take pains to distinguish themselves from the robot pundits. But, there is something so magical about the creation of artificial living creatures—mechanical entities with lifelike behavior—that even the soberest of the researchers find themselves wondering what lies ahead for their creations, and for humankind. The robot revolution will happen, whether we like it or not. From now on, we and the robots are in this together. All the more reason, thought Peter and I, to try to figure out what's coming down the pike.

Even before 1920, when the Czech writer Karel Čapek invented the word "robot" in his play *R.U.R.*, the world had begun to embrace the concept of artificial workers with humanlike capacities. Japanese inventors and artisans had created tea-serving automata, or *karakuri*, as early as the seventeenth century. Automata—mechanical contrivances designed to act as if they were under their own power—were familiar diversions in eighteenth-century European courts. And by the nineteenth century, automatons were creeping into science-based fiction and folklore in the form of golems, clockwork men, and Frankenstein's monster.

One of the first attempts to match reality to fantasy occurred after the Second World War, when aerospace engineer Joseph Engelberger (see page 186) conceived of machines that could perform repetitive tasks tirelessly and more accurately than their human counterparts, and brought robots to the factory floor. What began primarily as a field for mechanical engineers grew to include engineers of all stripes. Their machines were in essence puppets—expensive, beautifully designed devices that were controlled completely by the strings of their instructions. They couldn't think, create, or react; they simply performed their tasks, moving with the reflexive precision of a pendulum.

Robotics did not acquire its present scope until the arrival of modern computers, which inevitably brought with it the idea of stuffing some sort of brain into the robot. In the 1940s, English mathematician Alan Turing laid the groundwork for artificial intelligence, famously theorizing that a machine is intelligent when there is no discernible difference between its conversation and that of an intelligent person. At the Massachusetts

Sometimes described as the grand old man of Japanese robotics, Hirochika Inoue (left) of the University of Tokyo is one of the directors of the nation's massive effort to develop a humanoid robot.

Musing over the possible hierarchy of humans and humanoids, Takeo Kanade (above), the director of the Robotics Institute at Carnegie Mellon University in Pittsburgh and an expert in vision systems, posits, "We are facing the fact that we may not be, any longer, the single entity that does a better job in all aspects. How we as human beings will react to it, I don't know. But we are surpassed by many artificial things already. We don't mind that we have turned computing numbers over to machines; humans are not afraid of that at all."

Institute of Technology, trail-blazing computer scientists John McCarthy and Marvin Minsky founded what in the late 1950s became the world's first laboratory devoted to artificial intelligence. One goal was to use artificial intelligence to advance the study of human intelligence. Another was, of course, to build robots.

In those optimistic days, computer power was growing so fast that true artificial intelligence seemed to be just around the corner. It wasn't. Some AI researchers were able to program computers to behave intelligently in certain narrow functions, but they were never able to create a machine that could speak or read or solve unexpected puzzles. Dumping a dictionary into a computer—their approach, roughly speaking—didn't produce a book. Minsky tried to build an "intelligent" arm that could stack blocks atop one another. It never worked. Beset by difficulty, artificial intelligence as an active field of research declined in the 1980s.

Partly to blame is the inherent difficulty in creating a simulacrum of a phenomenon that nobody understands. If the nature of intelligence and consciousness remains a subject for speculation to this day, how can scientists manufacture it in artificial form? If we needed to know how our brain works—where thoughts come from and how memory works—in order to use it, all of us would be in a heap of trouble. Even the scientists who have charged themselves with the task of discovering the secrets of the brain, and are shrinking the pile of unanswered questions and conundrums at a faster and faster pace, are still working at the level of the educated guess. And if they get to the bottom of the pile of riddles, will they have the answers they seek? If the magic of a single thought is made not of illusion but allusion, will its genesis be any more possible to discern?

In 1980, Marc Raibert established the MIT Leg Lab, home to the first robots that dynamically mimic human walking—swinging like an inverted pendulum from step to step. Famously, Raibert even built a robot (above) that could flip itself in an aerial somersault and land on its feet. In 1993, he left the field to found Boston Dynamics Inc., in Cambridge, which translates his discoveries about humans and animals in motion, into animation.

Many roboticists today avoid the quagmires of AI by building what are in essence dumb machines, without a hint of consciousness but programmed cleverly enough to perform complex tasks—searching for breaks in a municipal sewer system or pumping gas at a service station (see page 195). Using cheap, scavenged electronic equipment, Mark Tilden, a researcher at Los Alamos National Laboratory (see page 117), can make small, insectlike machines that walk over irregular terrain with as much aplomb as if they had eyes to see where they were going and minds to adjust their step.

Does a robot need to have much of a brain? It depends on what you want it to do. Usually, machines we saw in the laboratories were bolted to a bench; some could maneuver around a finite pristine space under close supervision. But if robots are to inhabit the world's kitchens, as Tilden puts it, "You probably want the robot to know that it shouldn't suck up the cat kibble." (Let alone the cat.) When a robot operates in a human environment, the programming becomes more difficult. Safety concerns, mobility issues, and space requirements suddenly emerge. If more than one task is involved, the difficulty increases exponentially. Even a harmless, just-for-fun device like the Sony AIBO robot dog (see page 224) is subject to these constraints—that's why it moves slowly, is soft-edged, and costs twenty-five hundred dollars.

Sophisticated programming alone is not enough to make a machine seem lifelike. Curiously, what is required often is not great intelligence or startling skills, but randomness. Predictable behavior is computer-like; randomness is human. To accommodate this perception, Sony has added touches of spontaneity to the AIBO: in no particular or repeating order, at any given time, it might "play" with its ball or lie down and wave its legs in the air. But the notion of randomness in a machine dismays Engelberger, the robotics pioneer. "Maybe it is more fun and interesting if it screws up now and then and it does something a little different," he told me. "But I don't want that. I want the thing to be utterly reliable." ("Robots like the AIBO have a different purpose," I observe. "Yeah," he says. "To horse around.") It's understandable that Engelberger would feel that way—he designs industrial and health-service-oriented robots, which must be undeviatingly dependable. But even in the more relaxed atmosphere of the home, unexpected robotic behavior could be less than charming. A vacuum-cleaning robot that spontaneously broke out into a little dance might be funny the first few times, but a householder's patience might wear thin if that expensive robot vacuum cleaner were dancing instead of working, and wearing out its custom wheels and the carpet in the process. Thus even the most well-programmed, occasionally random automata will not be fit companions for people. If robots are to fulfill their creators' dreams, they will have to be given truly intelligent brains, which means that even if they want to, researchers will no longer be able to avoid wrestling with the riddles of AI.

Today, there are two main approaches to creating a clever machine: weak and strong artificial intelligence. Weak AI is the argument that a machine can simulate the behavior of human cognition, but it can't actually experience mental states itself. Even though such machines would be able to pass Turing's test of intelligence, they would still be little more than extra-complex clock radios. Proponents of strong AI argue, by contrast,

that machines are capable of cognitive mental states—that it is possible to build a self-aware machine with real emotions and consciousness.

Strong AI greatly distresses some philosophers, including John Searle of the University of California at Berkeley. If a computer can have cognitive mental states, he points out, then a human mind would have to be simply a computer program implemented in the brain. To Searle, this contention is absurd; consciousness is a first-person, subjective phenomenon that no mechanical computation, no matter how sophisticated, can produce.

Daniel C. Dennett, a philosopher at Tufts University, takes the opposite view. Consciousness, he says, is at its core algorithmic—that is, the brain has a series of rules for dealing with incoming sensory data, and the summed execution of these rules in the lower strata of the mind generates consciousness in the mind's upper strata. If Searle is right, robotics faces inherent limitations—we will never be able to build a truly intelligent machine. Dennett offers a more hopeful picture, at least for roboticists. But there is a chance that both might be wrong. Robots may need a brain to do everything their advocates imagine, but they may not need a brain that is humanlike. It is possible that circuitry utterly unlike the human brain could make robots behave in ways that seem indistinguishable from the workings of intelligent consciousness.

The construction of such machines—a race of intelligent aliens made right here on Earth—would be an ironic triumph for AI advocates. Proof that artificial intelligence is possible, these robots would still be incomprehensible; they would provide little or no insight into the human mind. Worse, they would plunge humankind into an immediate ethical quandary. If a robot has a brain equal to that of a human being, should it also have the legal and political rights of a human being?

The question of robotic rights could be less of a concern for the robots Kris Pister envisions. An electrical engineer at the University of California at Berkeley, Pister has the courage to think small. He and his colleagues are working toward a robotic future that is dominated by what he calls "Smart Dust"—autonomous robots no more than the size of a gnat (one cubic millimeter). Although equipped with sensors and ways to move around, each tiny machine will be relatively simple. But when thousands of them are combined into a network that permeates the environment, the totality will be capable of extraordinary things. Sprinkled on a child's clothing, the minute devices could monitor his or her location and state, sounding the alarm if the child climbed out of the crib or onto the back of a chair, or seemed to be choking. Instead of using a keyboard, people could stick Smart Dust on their fingers and run their computers by making finger motions. By scattering Smart Dust equipped with moisture and acidity sensors on food, homeowners could monitor the freshness of the items in their refrigerators. Mechanical motes on workers' clothing could signal office heat-

Rodney Brooks of MIT (with the latest incarnation of Cog, his humanoid robot) believes it likely that robots can achieve humanlike intelligence and consciousness. But when that happens, he says, it will be unethical to have them work for us—we shouldn't treat our creations as our slaves. He says, "You get into the moral question—would it be okay to breed a race of subhumans? Essentially, enslavers thought they were dealing with subhumans. If that's not okay, will it be okay to deliberately build subhuman machines? And certainly we feel now it's okay. We don't feel any empathy for the machines but that may be a consideration ultimately. I think we're a long way from having to face it ... but the landscape is going to be so unimaginable that it's hard to say sensible things."

Metallic flakes wafting from his hand, Kris Pister of the University of California at Berkeley demonstrates one possible offshoot of robotics research—Smart Dust. Miniature machines, each the size of a dust mote, may eventually saturate the environment, invisibly performing countless tasks.

ing and cooling systems to raise or lower the temperature. In effect, the dust would turn the entire environment into an invisible robot, constantly on the alert to do the human's bidding.

The first exemplars of Pister's version of a microelectrical and mechanical system are expected in 2001, but they will only be able to perform simple tasks like monitoring the temperature around them. Pister's students are developing minuscule legs and wings that will make the dust more mobile. Greater communication power and more complex sensors will come later. But the goal is always the same: individual components that can work together in clouds, literal clouds. Pister believes that in the future millions of these tiny robots will be constantly floating through the air. "When you fly across the country," he told me, "pilots are always saying, 'The last guy who passed through here 20 minutes ago told us that we are going to hit some turbulence.' If these guys were passing out Smart Dust from the back of the plane, they could query it as they went through to see exactly where the turbulence was."

Because inventions that don't yet exist can't be photographed, Peter set up a photo illustration to conceptualize Pister's ideas. The researcher stood before a projection of a pop-up micro Fresnel lens—the kind of lens used in lighthouses—and blew glitter into the light. I asked what he saw when he looked at the finished picture (above). "When I see the Fresnel lens," he said, "mentally I see a beam of light come out that's modulated on and off and transmitting information. Or maybe it's drawing a picture on a screen or on the back of my

retina. Then, the sparkling glitter coming out of my hands—when I see that, I think it's the little lasers that the Smart Dust particles use to communicate with each other. On our time scale, we see these things rushing out like a mad jumble. On their time scale, it's a very slow, beautiful, underwater dance. They're talking to one another as they're flying through the air, telling each other what they are seeing—'Who's where?' and, 'Let's set up our network.'"

Smart Dust has obvious military applications and in fact the research is backed by the US military's Defense Advanced Research Projects Agency (DARPA). Pister admits that his ideas could have a downside: permeating the world with invisibly small robots will make it possible to watch anyone, anywhere, at any time, but thinks the benefits of this technology will far outweigh the negative.

Without the Pentagon, US robotics in its present stage of evolution could very well collapse; in our research for this book, it was rare to find a top American researcher who was not sponsored at least in part by DARPA or the US Navy's Office of Naval Research (ONR), though other entities such as the National Institute of Health also provide some funding. Even foreign nationals like German researcher Frank Kirchner of the prestigious German National Research Center (see page 113) has US military funding for a robot project with Northeastern University's Joe Ayers (see page 110).

Is military backing for robotics a touchy subject with researchers? Not particularly. As aerospace pioneer Paul MacCready (see page 156) dryly observes, "It's been my experience that whatever we've done for the military has done more good for the nonmilitary." Generally, researchers go where the money is. DARPA's program managers solicit, vet, and fund proposals submitted by universities, private research laboratories, and corporations; they run conferences and meetings for prospective and current awardees just as professional organizations do for their members—an especially valuable resource in robotics, say researchers, because of the field's multidisciplinary nature. And the military supports nonsecret projects with humanitarian uses, including robots that search for and disarm land mines (see pages 101 and 111). "We'll play in anyone's backyard," says Georgia Institute of Technology's Ron Arkin of his lab's Pentagon support (see page 152). "But we'll only do open, unclassified research. That's where I draw the line, personally. Everything we do we can publish."

There is an unavoidable bottom line. "You have to have somebody who's a receiver for the technology," says Arkin, lamenting the lack of support in long-term robotics research by US corporations, which generally want a more immediate return on investment. "The question is," he says, "what industries are going to be ready to nurture that technology for the five or ten years it might take to build that product?"

Because its US-designed constitution is intended to forestall military adventurism, Japan does little defense research. Instead, its private corporations and the government are heavily involved in robotics research for commercial applications. Japan's equivalent of DARPA's robot program is its nationwide initiative to create a humanoid robot industry. The country's historic legacy in robotics has centered on *mekatoronikusu*, or mechatronics, a fusion of mechanical and electronic engineering, and it expects that its future will be chock full of autonomous walking robots of the humanoid persuasion. The country is in the midst of a ten-year national project to build just such a robot, which, says the University of Tokyo's Hirochika Inoue, "will walk in an unstructured environment and perform complex tasks." Japan's hope is that, among other applications, such a robot will fill the need for elder care in their aging population and situate the country at the forefront of a new humanoid robotics industry.

Inoue, a respected roboticist and a principal member of the project's governing committee, calls the humanoid project Japan's grand challenge. Though some researchers, such as Shigeo Hirose of the Tokyo Institute of Technology (see page 89), question such a robot's viability, Inoue is fiercely supportive of the project, which he sees as a technological lifeline for a country left behind when the software industry took off.

"We need to find that killer application for the humanoid robot," Inoue says, grinning at his purposeful slip into hackerspeak. "We will approach this very carefully, for we must define a good future for the humanoid robot." "For what type of business?" I ask. Inoue has ready answers: "A business network is needed. Maybe not a General Motors, but the General Humanoid Company. This company would design, develop, manufacture, and repair robots. Then there's the Humanoid Software Corporation, the Humanoid Taskware Corporation, and other technical support businesses. Also, humanoid options packages. Vision, hands. The Humanoid Tuneup Company. You can buy a service plan. These kinds of support companies will be needed to create a humanoid society."

Inoue's view of this peaceful coexistence between man and his servant machine is markedly different from the future supposed by those who either fear or desire the technological pursuit of atomic-level mechanization. Beyond the tiny microworld of Smart Dust, researchers are nurturing nanotechnology—the manipulation of materials on an atomic or molecular scale. Advances in miniaturization are expected to reach the level of individual atoms and manufacturing will be done to atomically precise specifications. Nanotech enthusiasts predict that creating molecule-sized machinery will be a world-changing technological and cultural milestone. Medicine, biotechnology, manufacturing, and information processing will forever be transformed. Endorsing the potential of nanotech, the United States announced a massive research initiative into molecular machines as the last century drew to a close.

Among the most common dreams for nanotech is to create super-small robots that can be injected into the human body. Tens of thousands of them will swim in our bloodstream, adding firepower to the immune system. Tirelessly seeking out viruses, tumor cells, and carcinogens, this network of nanobots will help us live longer, healthier lives. That's the optimistic view, of course. In Neal Stephenson's sardonic nanocentric novel *The Diamond Age*, rival clouds of Smart Dust wage aerial guerrilla warfare, covering the streets with a dust of minute, blackened machinery known as "toner." Sprayed like dandruff across people's shoulders, invisibly small nanobots monitor their every move on behalf of governments, corporations, private groups, and malicious individuals. Thieves are put to death with a syringe full of minute machinery; once injected into the bloodstream, the tiny engines swim around the body and, in response to a radio signal from an executioner, explode simultaneously. In this fictional scenario, robots have transformed daily life, but not necessarily for the better.

Such fascinating wild-eyed speculation is brought back to Earth by a story told to me by Marc Raibert (photo, page 24) whose odd-looking one-legged creatures sprang forward, hopped in the air, and somersaulted head over heels with a lightness and fluidity unlike any other Earthbound robots. Throughout the 1980s Raibert used his machines to explore theories about how living creatures moved—theories that roboticists have adapted for their own creations ever since. Although he left robotics and his directorship of the MIT Leg Lab as well in 1993, Raibert is still the researcher most often mentioned admiringly by other researchers both in the field and out. He retired from academia, in part, because of the first and only time that his creations left the laboratory and had to perform—on demand—in the real world.

"We were asked if our robots could be used in the Sean Connery movie *Rising Sun,*" Raibert explained. The movie featured a high-tech industrial plant in which the robots were supposed to operate. "My crew of three students and I set up on the first day and put things on a sound stage, so the director and Connery could see the robots. The robots ran just fine and they said, 'Oh, that's neat.' Then we had to go out on location, which was a water treatment plant. It was outdoors. It was hot. There was dirt and stuff on the ground.

"Our robots had only worked in the lab. It's one thing for the robots to work where you have a crew of students and technicians and nothing but time. All you have to do is report your findings. First there's development time, when you record data and run the things, and when something goes wrong, you fix it. And then you go again. Or you look to see what's happening and improve things. And then it's demo time—when you make video-tapes. The things only have to work for 40 or 50 seconds at a time—it doesn't matter which 40 or 50 seconds. Then you get your stuff together and make a presentation.

"We were at the set the first day that they were actually filming. We were out there fiddling around on the place that we came to know as the 'grassy knoll.' The film crew was 300 yards away with their camera, doing a big, wide panning shot of this futuristic-looking place. All of a sudden our walkie-talkie starts squawking and we hear, 'Roll camera, roll sound,' and they say, 'Start the robots.' For us to start the robots, we have to start our computers, boot them, get the gyroscopes up, uncage the gyros—all this stuff. They were already filming their shot! Then we had problems with the sensors and the robots getting full of dirt. The computers kept crashing because it was 100 degrees and they were used to being in the lab. We blasted open a hydraulic hose and the oil shot everywhere. It was not like we missed by this much [indicates inches]. We were nowhere in the ballpark.

"The final day we got some great stuff, which they used in the movie. It looked just perfect—just like the robots were running. But I went home and started to think about that and about the difference between making robots work in the lab and making any product, like the paper this business card is on, and doing it with a finite budget. And you have to take the results and put it in a cardboard box and ship it around the world and people take it out and it has to work." Raibert's own company, Boston Dynamics Inc., now uses his theories of motion to build artificial creatures. But these virtual creatures exist only on computer screens; they are computer models used by animators for films. This kind of robot, Raibert seems to be saying, is the only kind that is ready to leave the laboratory—a simulated robot with no physical existence. This won't always be the case.

The future is an amazingly fertile sphere for robot pundits, and the prognosis for mankind's future well-being ranges from pretty good to downright scary. Sun Microsystems cofounder and chief scientist Bill Joy is one of those who has come to believe that if mankind doesn't start practicing what is in essence safe science and technology, then biotech, nanotech, and robotics will triumph to the detriment of mankind, shaking us from what he seems to have assumed was the relative safety of our lofty perch atop the species hierarchy. What unnerved Joy, he related in a *Wired* magazine article and at a subsequent conference at Stanford University, was a chance conversation with John Searle and inventor-author Ray Kurzweil. Kurzweil, whose work in speech recognition is considered state-of-the-art, has an unusually good record for technological prognostica-tion. And when he told Joy that he thought mankind's rapid technological advances would soon blur the difference between man and machine, the notion scared Joy into action. In Kurzweil's telling, the advance of robotics was not the distant prospect assumed by many; instead, it sounded like an inevitability.

The thought of smart new machines helped prompt Joy to write a manifesto of sorts, proposing that human beings embrace a platform of technological stasis. "These technologies are just too much for a species at our level of sophistication," Joy said. His fears were not allayed by Searle's argument, by now well-known, that a machine cannot acquire consciousness. What dismayed Joy more than the possibility of the robotic master race of the future was what humans most probably would do to each other using the tech-nologies of the Information Age. "The threat to humanity in the last millennium—nuclear, biological, and chemical weapons of mass destruction—was largely confined to the military arena," Joy says. In the next millennium, *Homo sapiens* will be in much more danger, because of the democratization of technology. As

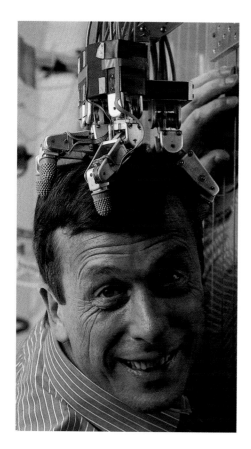

Once intelligent robots are created, argues Kevin Warwick of the University of Reading, it will not take them long to realize their superiority to flesh-and-blood beings. "The human race as we know it," he has written, "is very likely in its endgame"—machines will wipe us out. Believing that this dreadful scenario is unavoidable, Warwick nevertheless continues his research into robotics.

technology gets cheaper and easier, Joy believes, it will provide huge numbers of people—potentially every single person on the planet—with weapons of mass destruction. Like giving a gun to a baby? No. According to Joy's scenario, it's more like giving a deadly arsenal to a network of indiscriminate psychopathic killers.

Throughout most of the last century, only a few large, bureaucratic nations had weapons of mass destruction—until 1990, for instance, there were only five acknowledged nuclear powers. But with the growth of biotechnology, nanotechnology, and robotics, small groups—even individuals—will be capable of extraordinary destruction. The constraints that impede states are not necessarily felt by individuals. In a world with ten thousand or ten million people capable of wielding murderous nanobots, what are the odds that all of them will be sober and respectable world citizens? "We are," Joy said, "on a course of human destruction." The only way to avoid this threat, in his view, is "relinquishment." People must collectively ban certain kinds of technology tomorrow, as they banned genocide and slavery in the past.

The problem with relinquishment, noted Robert Wright, the author of *NonZero: The Logic of Human Destiny,* in a *Slate.com* essay, is it would require that certain Constitutional rights be "recalibrated." He argues, "For all I know, it's true that in 20 or 30 years these nanobots, by malicious design or by accident, will run so rampant that we'll be fondly reminiscing about the days of termites. On the other hand, this is basically the same problem that is posed by self-replicating biological agents [e.g., viruses]. In both cases we're faced with microscopic things that can be inconspicuously made and transported and, once unleashed, whether intentionally or accidentally, can keep on truckin'."

The best attack on viruses—and on many if not most of the other problems of humankind—is science, which depends on unfettered inquiry. If relinquishment mandates telling scientists what they can or cannot study, we may be making ourselves even worse off. Technological evolution is usually a double-edged sword. More than that, is relinquishment even practical? Given the difficulty we already face restraining material on the Internet, is it reasonable to assume that we could in fact rein in or wholly govern technology? At a Stanford University conference called "Will We Have Spiritual Robots in 2100?" Kurzweil acknowledged the downside of technology. Nanotechnology and robotics could be hugely important to his own work in speech-recognition hardware—microscopic nanobots in the body could internally scan the brain, providing further information to the software. But, he said, "It's not feasible to select the positive from the negative sides of technology.... The only way to forgo technology in general would be to have a totalitarian system that would basically use technology to enforce no one having technology. It isn't feasible to stop technology in specific areas that are going to be dangerous." Slamming on the brakes, he said, is not possible.

Though my background files were peppered with press accounts of his mildly outrageous exploits, cybernetics professor Kevin Warwick (photo, page 29) initially didn't strike me as all that different from other robot researchers I'd met. Sure, he'd had a chip imbedded under his skin once for a week-and-a-half, in an experiment that supposedly triggered doors and such to open at his approach, but our visit to his robotics lab outside London at the University of Reading passed uneventfully. He was between projects, and his lab robots, which included a large metal cat named Hissing Sid, were in various states of flux and disrepair.

Some months later, in Tokyo, a news brief caught my eye: an English gentleman had been refused a seat for his metal cat on a flight bound for Moscow. I knew immediately it was Hissing Sid and his keeper. After Warwick popped up as the cover boy on *Wired* magazine, talking about a new, more sophisticated implant that would be attached to his nerve fibers, I caught up with him again. Warwick was getting hundreds of letters and E-mails in response to the article, he said. "Who's writing?" I asked. "Neurosurgeons, students, office workers," he replied. "It is an amazing variety of people—lawyers, you name it—but extremely positive and supportive, which is fantastic." Warwick's view of the future is in the long term just as unsettling as Joy's. The

chip in his arm, he believes, would be but the smallest first step to a much larger melding of humankind and machine. Our machines will not wipe us out, in his view. Instead, we will become them, grafting our human consciousness into extraordinarily fast and durable machines. *Homo sapiens* will vanish as a biological species, replacing itself with a new race of cyborgs.

With his research team, which includes a neurosurgeon and sensory researchers, Warwick plans to connect a chip directly to nerve fibers in his arm. The chip will record the signals passing through the nerve and broadcast them on a tiny radio to a computer, which will record them on a hard drive. "The signals are not just an identifying signal," he explained. "We are transmitting whatever is on the nervous system at that time. Whatever is picked up on the nervous fibers will be transmitted to the computer. The computer is used just for storage. I think that is critical. Some people might ask, 'How does the computer understand the signals?' Well, that is not what we are doing at all; that is someone else's concern. Subsequently, the computer will play the signals back, so the only thing that needs to understand what the signals are is my body."

Ultimately, the hope is to go from storage to playback and from playback to networking. At first, he will simply move his fingers. The computer will record the impulses and feed them back at a later time. If all goes well, Warwick's fingers will move again in response to the stored memories. The next step will be to send those stored memories to other people. One of the first of these others will be Warwick's wife, who also has volunteered to

A founder of artificial-intelligence research, Marvin Minsky of MIT now believes the field has taken a wrong turn. Rather than trying to build costly, clumsy physical robots, he says, researchers should concentrate their efforts entirely on computer simulations —that's the key to unraveling the nature of intelligence.

have a chip implanted in her arm. Will she be able to feel what it is like for Warwick to move his fingers? Can we communicate to one human being what it is like to be in the body of another? How Malkovich.

Warwick plans to situate his wife and himself in different cities—Reading and New York—and send each person the other's sensations. "My level of success will be indicated by fairly basic things," he told me. "If I can move my finger and her finger moves; if I can feel pain and she feels pain; and vice versa. If I feel excited and she either feels that she is excited or feels excitement from me—I sort of feel that we have to get some results from this because we'll be sending signals that the brain is not just going to ignore." "But no one really knows what these signals will mean," I objected. "No," Warwick agreed. "I'm the true experimental engine. I can't see why people haven't done it before." Warwick plans to try to tap into some emotions. "If we can record three or four different emotional situations and press a button to play one back, I will report what I am feeling. If you press 'Fear,' and I am reporting excitement, then obviously we are sort of off the ball a little bit. But it might be a nice little experiment. Press 'Fear,' and I get fear; press 'Excitement,' and I feel excitement. That would be utterly fantastic, to be honest. I think it would shatter the world." For the first time ever, one person would be able to feel what it is like to walk in another person's shoes.

"These experiments," he said, "tie in nicely with my own beliefs to where we can go and what we can do with it." In Warwick's view of the future, the world will be populated by hybrid creatures—part living flesh, part electronic device. Linking humans to machines, he said, involves "creating cyborgs—there is no way around it—people that are part human, part machine." Such entities, he admitted, will have "values that are very different from human values. But that is the way that we are going. I know that is a bit sci-fi-ish, but that is the direction. That is clear." He adds, "The destruction of the human race as we know it seems inevitable." It's startling to speak with technologists who hurtle full steam ahead in their research while predicting a bleak future for mankind because of it.

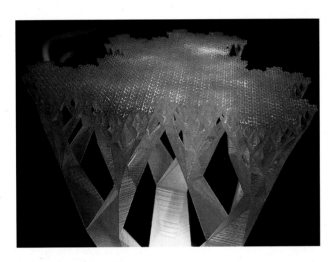

Dismissing the fears that robots will come to dominate their creators, Hans Moravec (right) of Carnegie Mellon argues that humans will literally become robots, "uploading" their consciousness and memories into their computers. Advanced manufacturing techniques, he says, will enable the creation of machines that will far surpass the dexterity of conventional mechanical manipulators and even human hands. Equipped with molecule-sized "nano-fingers" (above, in model), these devices will be able to create any physical structure, atom by atom.

In some sense, this is nothing new. Watching the first atomic blasts in the southwestern desert, J.R. Oppenheimer said that sin had come to physics. But Oppenheimer's group at Los Alamos was building atomic weapons to defeat Hitler. Warwick is plunging ahead with research that he thinks will contribute to the end of the human race in the name of ... well, what? One is tempted to argue that he must not believe what he is saying.

But Warwick is not the only robot pundit who both espouses a bleak future for mankind and works to bring that future closer with his own research. There's Hugo de Garis, a computer scientist who is building his own brand of artificial neural networks with the belief that he will one day be known as the Father of the Artificial Brain. In our conversation, de Garis (photo, page 28) whispered conspiratorially that though his work will help create artificially intelligent beings, he feels a moral obiligation to "raise the alarm" about the fruits of his research. Meanwhile, he was peddling a screenplay to Hollywood in which Terrans (real people) and Artilects (artificially intelligent beings) battle it out for the top spot on the food chain.

A slightly more benign view of the end of humanity is propounded by Hans Moravec (right), a robotics and AI researcher at Carnegie Mellon University. Today, Moravec says, people are capable of much more than any computer program. There is a lot of rich interconnected processing going on inside the brain—Moravec concedes the point. But he believes that robot intelligence will catch up to human intelligence and maybe surpass it by 2050. The prediction is staggering, because the cleverest robots today have the intelligence of insects. If Moravec is correct, robots will do the equivalent of evolving from ants to humans in less than a single human lifespan. These über-robots, in Moravec's view, will be the offspring of humankind. Except that unlike most offspring, they will not gradually replace us. Instead, we will become them. We will launch huge clouds of robots into space, he says. They will circle the stars, feeding off solar energy. But humans will not stay behind. Instead, we will convert our minds into digital form and upload ourselves into the clouds. Living for eons, our unrecognizable descendants will percolate the galaxy as digital entities: *Robo sapiens.*

Will our machines destroy us, make us immortal, or change us beyond recognition? There's no shortage of answers from *Those Who Know.* Warwick, Joy, Moravec, Pister, Minsky, and their fellows speak with utter certainty about the shape of the millennia to come. But the only predictable aspect of the future is its unpredictability. In a world where the brightest minds failed to anticipate the collapse of the Berlin Wall or the rise of e-commerce, what chance do we have to imagine the faraway contours of the robotic future? What is possible is to bear these various images in mind as warnings or promises while we pay close attention to what is happening today. And that is what Peter and I tried to do as we visited the laboratories of the scientists and researchers around the world who are creating the future envisioned by *Those Who Know.*

Robo sapiens

Ten years and tens of millions of dollars in the making, the Honda P3 strides down its course at the car company's secret research facility on the outskirts of Tokyo.

Body Electric

When people hear the term "robot," they think of a machine in a humanoid shape—a moving metal replica of themselves. From the benign figures in Isaac Asimov novels to the implacable assassins in the *Terminator* movies, the popular image of a robot has always involved two legs, two arms, and a head with eyes, ears, and a mouth. Yet professional roboticists almost always mean something broader by the term (although they don't always agree on exactly what).

A robot, many of them say, is any machine intelligent enough to direct its own motion and respond to its environment, no matter what its shape. Part of the reason for the disparity is that—as roboticists have learned—building a humanoid robot is an exceptionally difficult task. Like finding the cure for cancer, building a humanoid robot is a scientific Grail that scientists have futilely tried to reach for decades. Only in recent years have researchers begun to make rapid progress.

One center in the long quest to make a human-like robot is Waseda University in Tokyo, where the late Ichiro Kato and his mechanical-engineering students constructed the world's first life-sized humanoid robot in 1973. Wabot 1 (WAseda roBOT), as it was called, was a clunky metal being outfitted with rudimentary vision and speech synthesis. Wabot 2—Kato's organ-playing robot—achieved fame by playing for the thousands of people who filed through the Tsukuba Science Exposition of 1985, fueling the imagination of future generations.

Waseda acquired some serious competition in the next year, when Honda Motor Company began a secret multimillion-dollar program to build a walking robot with a human form. Unveiled in 1996, the Honda humanoid robot (see page 43) attracted so much attention that it pushed the Japanese government into launching a five-year, multimillion-dollar humanoid robot project.

Kato could not take part—he died in 1994. His robots are dusted off occasionally to give historical perspective to robot exhibitions, but his living legacy is the curiosity he nurtured in his students, who continue to work toward the goal he dreamed of— the full-sized humanoid robot. Waseda University has become a major part of the project.

One of those former students, Atsuo Takanishi, leads the Waseda team in a cramped lab that looks like a dormitory—beds and desks are packed end to end in the space. "My big excuse for this mess is that we are doing an experiment with a robot rat and a real rat," the professor explains when we visit. Because rats are active in the night, his students must staff the lab round the clock—hence the

presence of the beds. "Actually," Takanishi jokes, "my students just live in the lab." It doesn't look far from the truth.

Takanishi works only on humanoid or human-inspired machines. We see the crown jewel, his life-sized, walking humanoid robot, WABIAN-RII, for the first time in a video—it's walking and holding a box of candy in its hands. But we don't see it move in the (so-called) flesh until later, when we visit as his students are prepping it for a public exhibition. It kicks slightly from side to side as it takes one step and then another, moving five steps before it stops. Arms outstretched, four students hover about the machine like a gymnast's spotters as it goes through its short routine. When WABIAN stops, they struggle to haul the heavy robot back to its starting position.

Faith: Your university has been working on robotics longer than any other.
Atsuo Takanishi: We have been doing research on biped robots for more than 30 years. Professor [Ichiro] Kato started that kind of research. He was one of the first researchers in the world to research a robot walking with two legs. But it is very difficult to find a big market for that kind of robot, so we regard our research on the biped robot to be scientific research on human walking.

There is a lot of research into human walking in the medical field, but it always lacks enough quantitative information. [For example,] it is very difficult to measure stress because you would need to have some surgery to put a sensor onto the bone. A human has a lot of bones and the lower limbs have 246 muscles, so it's almost impossible to measure all the different pieces of human locomotion.
And so you are using robotics to research this.
Our research focuses on how a human walks. We are using robotics to do this—it's a kind of synthetic way to explain how a human walks. We mimic human design with the robot. The long-range purpose of this robot is to research emotion and intelligence. We think that the basis of intelligence is emotion—the most primitive emotions. And to research emotion, the body's hardware—the robot's hardware—is a necessary part.
You and other roboticists often bring up Ichiro Kato's name. What was his influence on you?
About 15 years ago, when I was a PhD student at Ichiro Kato's laboratory, there was a summer laboratory meeting where all the students had to present their research in front of several well-known robotics researchers who had graduated from Professor Kato's laboratory. Their comments were very severe to all the students. I introduced a new idea for walking control and we had a heated argument. Everyone in the audience except Professor Kato opposed my idea. Only he understood and supported me. I was

Robo Specs

Name: WABIAN-RII

Origin of name: Acronym for Waseda BIpedal humANoid

Purpose: 1) To clarify the motion-control mechanism of the human form from the viewpoint of robotics; 2) to establish a base technology to build the personal robots of the future.

Creative inspiration: The work of the late Ichiro Kato, often called the father of Japanese robotics research.

Height: 1.83 m

Length: 70 cm (across the shoulders)

Weight: 127 kg

Vision: 2 color digital cameras

Sensors: 27 rotary encoders, 2 6-axis force/torque sensors, a 3-axis gyrometer

Frame composition: Aluminum alloy, super-Duralmin, carbon fiber

Batteries: None

External power: Single-phase 100 V AC, 3-phase 200 V AC

KLOC: 10

Cost: $370,000 (components)

Project status: Ongoing—working on third-generation WABIAN robot.

Information from: Atsuo Takanishi

Looming out of the shadows in the Humanoid Research Lab of Tokyo's Waseda University, WABIAN-RII (left) is capable of walking and even dancing. A completely humanoid figure with two legs, two arms, two hands, and two eyes, it exemplifies the belief of many roboticists that their creations must adopt a human form to function in a world designed for humans—and that robots won't interact successfully with people unless they look somewhat like them. WABIAN sways from side to side as it walks, but its builders are not discouraged by its imperfections: walking in a straight line, which humans can do without thinking, in fact requires coordinated movements of such fantastic complexity that researchers are pleased if their creations can walk at all. Indeed, researchers built the robot partly to help themselves understand the physics of locomotion. It took decades of work to bring WABIAN to its present state: its first ancestor was built in 1972.

At a robotics exhibition in Tokyo, Samuel Setiawan (above, in white shirt) and two other Waseda University graduate students cautiously stand by during a lengthy prewalk checklist for WABIAN-RII. To their dismay, the robot initially has trouble negotiating the wooden floor, which is much springier than the concrete floor in the lab, where it had been programmed to walk. After some frantic reprogramming, Setiawan, the primary student researcher on the project, is able to make WABIAN walk its assigned path. Even so, the students have to spot the machine constantly. The students' care illustrates one of the chief worries of scientists trying to make robots for the home—their heavy creations might end up hurting the people they are supposed to serve.

able to continue and extend the idea, and by the end of that summer I had developed a new walking control method that could handle virtually any legged robot with a body or a trunk. Now that control method has been improved and applied to WABIAN and even to other robots in the Japanese humanoid-research project. That is, neither WABIAN nor I would be here if Professor Kato had opposed me.

What is the control mechanism that you decided to use?
The important thing is what we call the zero-moment point. If the zero-moment point is within the support polygon—

A zero-moment point is the point where all the forces are in balance. But what is a support polygon? Is that the area bordered by its feet? The area it is using to support itself?

For example, if the robot is standing still on one foot, the support polygon is exactly the same as the shape of the foot. My method is based on constantly controlling the zero-moment point, which is kind of a dynamic center of gravity.

How does this compare to the technology in the Honda P3?
They are using the zero-moment point concept exactly the same as we. This was the work of a Yugoslavian researcher, Miomir Vukobratovich. He invented the ZMP concept and his close friend, Professor Ichiro Kato, applied it to his biped walking robots.

Professor Kato published many papers about biped walking robots using the zero-moment point. The control method was developed so the robot could change its foot positioning when disturbed, such as if it were pushed by someone while walking. The Honda people are using very similar technology to control their robot.

So they are building on past research, but they certainly have more money and time invested in making a walking robot.
Yes, that is the point. They are very good at making hardware. They can invest so much money and so many engineers in one robot. In a sense, they are an integrator of research, gathering a lot of

information from the academic arena and many other research fields.

I know you ultimately hope to have robots like this interacting with people in their homes. What issues should be considered in the construction of robots for the human environment?

For this kind of robot, safety is a very big issue. In robotics, there has been almost no safety research. Industry keeps robots and humans separate to guarantee human safety. Because personal robots are always in contact with humans, I believe it will take a longer time than we expected—maybe twenty years. But we could construct humanoid robots in another way, as small humanoids, even pets.

Safety is not a concern with AIBO [the Sony robot pet]. I believe that within a few years we will have a lot more pet or entertainment-type robots around us. That will be the first step toward real personal robots. At the same time there will be research on robot safety. And eventually we will see real personal robots walking around us.

Let's change tack. Who do you really respect in robotics?

Personally, Professor Ichiro Kato, because he started robotic research in Japan. [He pauses for a moment and thinks about this.] As a mechanical engineer, I also have much respect for Professor Shigeo Hirose [see page 89], who has been developing fantastic robot mechanisms for years, and Professor Rodney Brooks [see page 58], who has been doing research on intelligent robotics, and Marc Raibert [see page 28], who used to make a lot of impressive running robots. He is a very, very smart man. I also have a great respect for Professor Paolo Dario in Pisa, Italy. He is developing medical and welfare robots that will play extremely important roles in our future society, supporting handicapped or elderly people.

Many researchers have told us that they respect Marc Raibert. He has his own commercial animation studio now, called BDI. Do you miss having him in the robotics field?

Yes, but I think he likes computer graphics much more than making robots. I think he is enjoying running his company.

Why do researchers respect him?

Marc Raibert is not American in personality. He says nothing more or less than what something is. He is a very honest person—very kind, always serious in his response to questions and is eager to understand.

In terms of where robotics is heading, are there pitfalls to avoid? Is anything happening now that you think is moving in the totally wrong direction?

There are some people who think we should construct the humanoid in a biological way [by manipulating DNA to create a robot]. I am not sure. I am not a biologist—but the human body is not totally

analyzed, so there are many unknown parts in biological systems. And some DNA treatments are very dangerous.

Machines that are constructed of living materials—

Traditional mechanical engineering has more than 300 years of history. So many basic problems are already solved. Now we can launch rockets to the moon or even Mars—it is a mature technology. We can use it to make a mechanical humanoid robot.

We know what will happen when we design the robot because there are many ways, using theoretical mechanics, to numerically confirm it. And we can use computers to simulate what will happen in the future. But using biology seems to me to be very dangerous; biotechnology has a very short history. So I always say that we should make a humanoid robot in a mechanical way.

Does it scare you that someone might create this biological new life?

That is very, very scary.

One of the leading researchers at Waseda University's long-term robotics project, mechanical engineer Atsuo Takanishi (above) studied under the late Ichiro Kato, a robotics pioneer—and superb fundraiser—who made the school into the epicenter of the field. Continuing Kato's emphasis on "biomechatronics"—replicating the functions of animals with machines—Takanishi now supervises the research group that produced WABIAN-RII (behind him in photograph). The latest step in the laboratory's long evolutionary struggle to create a robot that can walk naturally, WABIAN was funded in part by Japan's New Energy and Industrial Technology Development Organization as part of a nationwide humanoid-robot project directed by the government Agency of Industrial Science and Technology.

In a situation all too familiar to robotics researchers, Atsuo Takanishi (on right) is trying to make his creation work. His research team's robot, WE-3RIII (Waseda Eye Number 3 Refined Version III)—can follow a light with its digital-camera eyes, moving its head if needed. In the laboratory the robot worked perfectly, its movements almost disconcertingly lifelike. But while being installed at a robot exhibit in Tokyo, WE-3RIII inexplicably and violently threw back its head, tearing apart its own wiring. Now Takanishi and one of his students are puzzling over the problem—and will solve it only in the early hours of the morning before the exhibit opened. Back in the lab during the following months, the robot head swelled with sophistication. Aided by mechanical lungs, it can now smell and even blush.

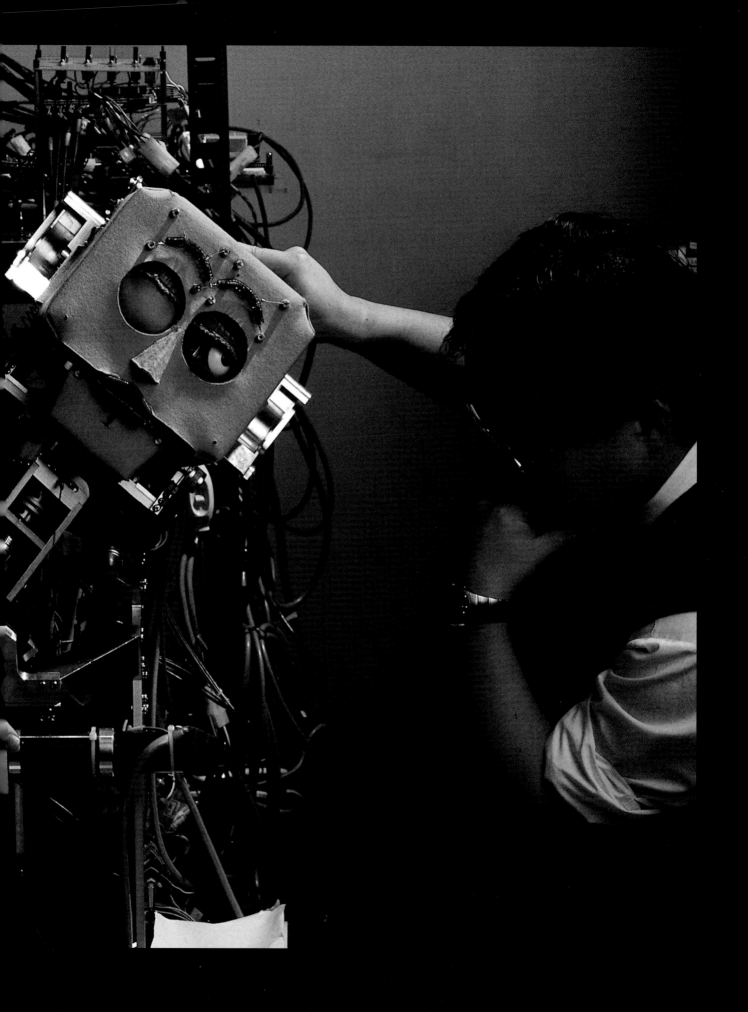

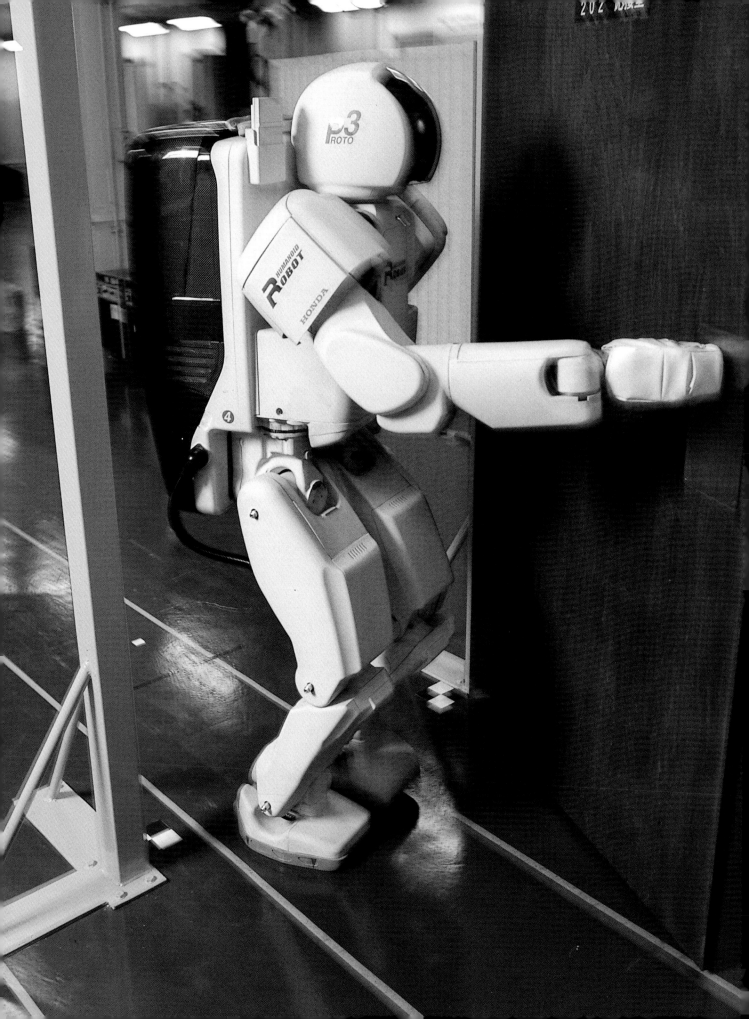

Million-Dollar Legs

The robot will start here, a supervisor says. "Here" is a strip of white tape on the floor of the Honda Wako Research and Development Laboratory, the highly secret robotics facility outside Tokyo run by the Honda Motor Company. For ten years, the research team operated in complete secrecy; outside the lab, only the very top brass at Honda knew that the company was spending millions of dollars to create a life-sized humanoid robot that could walk on two legs, like a human being.

When Honda researchers revealed their first prototype in 1996, the announcement stunned the robotics world, not only because the project was so successful, but because it was so successfully kept a secret. Even today the company is reluctant to allow outsiders into the research lab, but Peter is persistent and Honda finally agrees to let us see the latest version of their robot walk in the laboratory.

We are escorted to a large, windowless robot laboratory where there are half a dozen shrouded figures hanging from the rafters at various workstations. After the current version, Honda P3 (for prototype three), is activated, the supervisor explains that the robot will walk to a door and open it, pass through the doorway, close the door, and walk to the bottom of a staircase. This will be the end of Program One. After technicians snap on a safety chain, Honda P3 will tackle Program Two. Snugly tethered, the machine will attempt the difficult—and, given the possibility of a fall, potentially costly—process of going up and down a flight of stairs.

"Can the robot stop anywhere, if I want it to stop?" Peter asks. "No, no," says the supervisor. "Once it starts the first stage, it cannot stop until the end of its program." The robot is not controlled by operators but does follow preprogrammed routines.

Looking more than a little like a short, white-suited spaceman, P3 hangs just above the floor from a chain attached to its head. The robot's hands are in glovelike casings. "Humanoid Robot" is written out, in English, on its chest. The chain moves forward on a pulley and lowers the robot to the floor—knees bent, feet ready to be planted flat on the ground. A hard-hatted technician steps behind it to position the robot at the tape mark as it touches down. The chain moves away and the robot stands independently—in itself an impressive feat.

As the robot straightens, it begins to move, emitting faint grinding sounds. It moves toward the door, and its footsteps sound surprisingly wet, like the slapping of rubber soles on a linoleum floor. As a human might, the robot twists slightly at its torso to maintain the correct walking gait. When the robot reaches the door, it turns the knob and pushes the door open; the action seems perfectly natural. So natural, in fact, that when P3 turns to avoid the door before closing it, I can't help wondering aloud whether I'm watching a person inside a white plastic suit. I'm quickly assured by our hosts that I'm not.

A crowd of 15 or 20 white-uniformed lab technicians and engineers is watching the robot go through its paces. Their work has come to a standstill as they're not allowed to work when visitors are present. They have draped the robot bodies on their workstations with cloth to guard against prying eyes. Ours.

Program One ends without a hitch and the safety chain is attached for Program Two. The robot does truly keep its balance as it moves up the stairs. It turns at the top, bends slightly at the knees, then steps down until it reaches the bottom. It comes to a halt at the program's end.

We thank the technicians, who are huddled around a nearby computer monitor, grinning widely. I realize why and wave into the face of the robot. The technicians quickly look away . They're watching what the robot is seeing through its visor—Peter and me working in front of the robot—and they're a bit embarrassed at being caught. I'm embarrassed because I didn't realize earlier what they were doing. Peter wants to hang out with the robot some more but we have already overstayed our allotted five minutes by a half-hour. A few months later, we catch up with the Honda robot program's senior engineer, Masato Hirose:

Faith: How would you explain the Honda P3 robot to someone who has never seen it before and has no experience with robots?
Masato Hirose: When we began working, people at Waseda University had already built a humanoid robot that walked. So we asked ourselves, what could we do? And we thought, if we could develop a robot that walks at a speed equivalent to that of a human, it would be unprecedented. And if we could develop a robot that walks on two feet and is actually useful, that would be better.
Useful in what way?
For example, we may be able to make this robot walk on a flat surface in a large area, like a gymnasium. The robot could be operated, like playing a game.
You mean, for entertainment?
Yes.
Could people use it for something other than amusement?
Well, when we were still envisioning the kind of robot we wanted to develop, we dreamed of a robot that would be useful at home. But that is still a dream for us, to be realized in the future—our cur-

Robo Specs

Name: Honda P3

Origin of name: Prototype 3

Purpose: To cultivate a new dimension in mobility as an automobile company.

Height: 1.6 m

Weight: 130 kg

Maximum walking speed: 2 km/h

Vision: Digital video cameras

Sensors: Gyrometer, G-sensors, 6-axis force sensors on wrists and feet

Frame composition: We are regretful that some [questions] cannot be answered due to our PR policy.

Batteries (type, duration): Ni-Zn batteries, 25 mins

External power: DC 136 V

KLOC: We regret that some [questions] cannot be answered due to our PR policy.

Cost: We regret that some [questions] cannot be answered due to our PR policy.

Project status: Ongoing

Information from: Noriko Okamoto

Deftly opening a door, the Honda P3 (left) walks its assigned path at the Honda Research Center, outside Tokyo. The product of a costly decade-long effort, the Honda robotic project was only released from its shroud of corporate secrecy in 1996. In a carefully choreographed performance, P3 walks a line, opens a door, turns a corner, and, after a safety chain is attached, climbs a flight of stairs. Despite its mechanical sophistication, it can't respond to its environment. If people were to step in its way, the burly robot would knock them down without noticing them. Ultimately, of course, Honda researchers hope to change that. But, in what seems an attempt to hedge the company's bet, P3 senior engineer Masato Hirose is also working on sending the robot to places where it cannot possibly injure anyone. In the future, he hopes, the robot will be rocketed to distant planets, to remotely explore places human beings cannot yet visit.

Utterly ignoring the safety chain attached to the base of its "neck," the Honda P3 (above) confidently walks down a flight of steps in the company lab. More than a decade ago, at the beginning of the Honda project, the research team concluded that their robot would have to be able to walk, rather than simply roll on wheels. Wheeled robots, they decided, just couldn't function in a contemporary home full of stairs, toy-strewn floors, thick pile rugs, and other obstacles. Today P3 can walk with impressive smoothness. The only real sign of its robotic nature is the way it begins to walk with a little knee-dip, to compensate for the absence of a pelvis.

rent robot is not capable of that. We still need to reduce its weight, so it is equivalent to that of a human. There are also safety issues to be resolved. [P3 weighs 130 kg, and such a heavy object moving about a home could be dangerous.] But what we envisioned was—if we could develop a robot able to serve a human master, that would be very interesting. For example, when the master goes shopping, if the robot could accompany him or her, carrying the heavy bags, that would be very nice. In fact, this was the kind of thing that we envisioned at the initial stage [of development].

How far in the future is a shopping robot?
[Laughs]. It may take another ten years.

Only ten years! Japanese sidewalks are already so crowded, it's staggering to think what they would be like if everyone were accompanied by a robot. Maybe the robots could go shopping by themselves.
We're still at a stage where people have to develop their own sense about whether a robot could be useful to them. They could treat such a robot as a sort of mascot. I think whether these robots should come

into our daily lives or should remain in a special separate world [as in contemporary factories, where they are generally fenced off from human employees] still needs to be discussed further. We haven't reached any conclusions yet.

What about in a situation like taking care of the elderly? That's a concern in Japan, with the population getting steadily older.
Well, we're not at a stage yet where we can develop a robot that could be involved directly in taking care of people or nursing people. We haven't reached a stage yet where robots can treat or serve human beings.

Will that be P4?
No, not yet [laughs]. We still have a long way to go.

Maybe P10?
[Laughter].

What would you say to researchers who claim the Honda robot is old technology? Everyone in robotics is interested in it, but they also tell us that it's a lot of money for something built with old technology.
Well, actually we don't have any opinion about this. We just want to develop what we want to develop. I think that other people can say whatever they want about the technology.

When you look at the robot walk, after putting all this effort into it, what runs through your mind?
When I saw the robot walking for the first time, I had a deep sense of appreciation for my company. Because, first of all, the company allowed me to continue with this research and development. And as the saying goes, continuance is strength. I think that in our world today, very few profit-making companies would allow us to continue for twelve or thirteen years in the type of research where there's no guarantee that it's really going to work.

Why is a car company building a humanoid robot?
It's just something we view as another kind of mobility. An automobile gives you two-dimensional mobility. Three-dimensional mobility is an aircraft. And I believe four-dimensional mobility would be a robot that is a double of yourself.

When you think of ninjas, there is one technique that they supposedly have—where they can make an identical self appear. Several identical selves, even. It's something like that. You know, you might laugh when I say this but, for example, consider myself. I'm married. Let's say that because of work I was transferred somewhere away from my wife and my home. Maybe this robot could be with me, and my wife would remotely control this robot to monitor what I'm doing—whether I'm doing anything improper or whether I'm not cleaning up my room.

Robotically monitoring errant husbands? I like that! So why did you decide, out of all the alternatives, to build one in human form? Bipedal walking, for instance, is much harder than getting something to move on wheels or tracks. And presumably, a shop-

ping robot could wheel along just as usefully as one that walked.

We had three alternatives. A robot with a maneuvering movement like [Shigeo] Hirose's snake robot [see page 89]. The second alternative was a robot that would move on legs, like a horse, and the third alternative was something that moves on wheels, like a car. As you know, Honda was already involved in machines with four wheels, so we eliminated that alternative. Then we thought that a robot that would walk on feet with legs would be the most interesting. And this kind of robot would also have the capability to walk in the mountains.

You want the robot to be able to navigate rough terrain?

Yes, but we still don't fully understand, technically, how far this robot can walk like a human. When a human being climbs a mountain, he doesn't use only his feet or legs. Of course not—he needs to use his hands, his arms. It's a synchronized movement of feet, legs, hands, and arms.

The current P3 can move its hands and arms and feet and legs together—we call this synchronized control—but the movement is limited. For example, let's say there's something heavy in front of a table and the robot is supposed to lift it onto the table. When it carries or lifts the object, its hips go backward automatically, to counterbalance the load. And when the hips are moved backward, in order to keep its balance, the robot has to use its hands and arms and legs so that the center of gravity will remain at the soles of its feet. That's all that this current model can do.

Is the upper body tele-operated? When we saw the robot walking, was its upper body controlled by operators at the computer console?

Not everything is tele-operated. For example, as I explained to you, when the robot lifts something, it tries to maintain its balance by itself. In the same way, when the robot tries to open a door by using the doorknob, it can sense the reaction coming back from the doorknob. And there's also the force that is applied when pushing the door open. So in that sense, it's not tele-operated. It's maintaining its balance by itself.

Clearly, that requires some computer power. How would you characterize P3's intelligence?

Well first of all, this P3 does not have the type of brain function that a human being has. It doesn't have the capability to think, to plan, to develop strategies, like a human being can. Rather, it's more focused on movement. Like moving its feet and moving its legs, and trying to maintain its balance when lifting objects. It's focused on the equivalent of a human being's spinal cord. So I don't really know whether you can call this intelligence.

Different roboticists have different definitions for

machine intelligence. An industrial robot that's painting a car has a certain limited amount of intelligence. It has only one task. It does its task and does it very well generally.

I agree. I don't know whether there is such a term but I think it's more like movement intelligence.

Why do you call the robot Prototype 3 instead of giving it a more personal name, as many robot builders do?

The reason we just call this P3, for Prototype Three, is that I personally don't like the idea that a robot can reflect the will or personality associated with a name. Often we are asked, "Is P3 a male or female?" But I always say it's neither. For me, a robot is a machine. It's an object and we are still in the process of research and development. And so I do not like to

regard this robot as being comparable to a human being and needing a name.

For many years, the Honda robot project was unknown to anybody outside the robot team here. It's really hard to keep a secret, even for a little while.

How did you keep this project a secret for so long?

Well, yes, you're right. This project was maintained as a secret for over ten years within our company. Even within our company only the top management knew about this. We did not issue any papers during this time, but I think that we were able to keep it a secret because we were working in a room without windows [laughter].

Was your wife in on the secret?

Well, actually, I wasn't supposed to tell her but I did [laughs sheepishly]. During the first years, I wasn't supposed to tell anyone about this. But I was undergoing so much struggle and difficulty, so many obstacles, and nothing was working. So, I had to talk to someone. I talked to my wife because I really needed to let things out. I think my company will forgive me.

Lights from futuristic concept cars reflecting in the shiny column behind his head, Honda P3 chief engineer Masato Hirose (above) has been entrusted with the transportation company's hopes of getting beyond wheels. Honda began its humanoid-robot program in 1986, intending to develop not the static, special-purpose robot common in automobile factories but a mobile, general-purpose robot that could be useful in the household. Hirose and his co-workers won't talk about it, but it seems clear they had no idea at the outset that it would take them more than a decade to create a robot that could walk. And they are still far from one that is ready for household use.

Baby's First Steps

Robo Specs

Name: None

Purpose: Test bed to check novel genetic algorithms and evolutionary programming.

Creative inspiration: In the future, robots will work in human areas, such as the home. It will be very convenient for them to have the same locomotion mechanism as humans. So a humanoid robot is desirable and necessary. Our robot is the first step toward it.

Height: 1.06 m

Weight: 24 kg

Vision: None

Sensors: 2 6-degree-of-freedom force-moment sensors

Frame composition: Aluminum

Batteries: None

External power: 100 V AC

KLOC: 25

Person-hours to develop software: 4 yrs

Cost: $55,500

Project status: Ongoing

Information from: Yasuhisa Hasegawa

Lurching from side to side like an infant figuring out how to walk, the biped-locomotion robot (right) in the Fukuda Lab at Nagoya University tentatively steps forward under the parental supervision of graduate student Kazuo Takahashi. Designed by Toshio Fukuda, a professor of mechanical engineering, the robot is intended to test what Fukuda calls "hierarchical evolutionary algorithms"—software that repeats an action, learning from its mistakes until it approaches perfection. It's a slow process: after months of work, the machine still isn't able to manage its arms, which lie forlornly on the cabinet behind. If Fukuda's team—or any other research group—succeeds in creating a machine that can teach itself to walk, many roboticists believe that they will be one step closer to understanding the same capacity in the human mind.

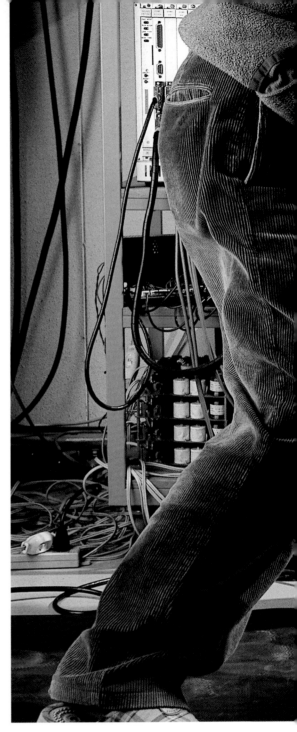

A**lthough the city of Nagoya is dominated by an enormous concrete replica of an old castle, it is in fact a muscular, modern, industrial place, housing Japan's biggest automobile factories— the Detroit of Japan. Sweeping us through his laboratory at Nagoya University, Toshio Fukuda is as energetic as the city he works in. A professor at the Center for Cooperative Research in Advanced Science and Technology, he is also president of the Robotics and Automation Society of the Institute of Electrical and Electronics Engineers, an influential professional association.

When we arrive at the school, the halls are crammed with bleary-eyed graduate students in coats and ties—it is the day of oral examination, which requires formal attire, but they have been up all night studying. Forceful and charismatic, Fukuda cuts through the mob of students like a cowboy cutting a herd of heifers, shouting out encouraging words as he moves. Eventually, he spies his target, a junior professor, who will escort us to a basement workroom.

In the room is a laboratory with a bipedal, walking robot that is one of Fukuda's newest projects. Its principal builder, grad student Kazuo Takahashi, blanches a bit when we arrive—he is being asked to demonstrate an obviously incomplete robot before one of his professors and a couple of foreign journalists. The robot—it is unnamed, because Fukuda doesn't like the somewhat standard practice of giving names to machines—is in the middle of the floor, surrounded by castoff pieces. A power cord hangs from its back like a long plastic tail. Down a staircase comes another grad student to help. The second student operates the keyboard while Takahashi hovers over the child-sized robot like a proud father monitoring a baby's first steps.

The metaphor is proper, because the machine walks much like an infant who is just learning how to put one foot in front of the other. Its movements are surprisingly quiet—I barely hear the whir of its actuators. Takahashi is beaming with a mixture of pride and relief.

Afterward, Fukuda invites us to a party celebrating the end of exams. Unlike the rubber-chicken fodder at most big US universities we've been to, the food at Nagoya is delicious—plates of sushi. The academic predilection for speeches is the same, though. When we ask Fukuda about general subjects like the future of robotics, he's delighted to share what he thinks. But when we ask him about specific subjects such as this walking robot, he's secretive about the details—charmingly secretive, but secretive nonetheless. Not only won't he tell us whether it is part of the Japanese national humanoid-robot project, he won't even tell us why he won't tell us.

Faith: What do you think are the main directions of current robotics research?
Toshio Fukuda: Robotics will go in three directions. First, more autonomous and intelligent robots for advanced missions, such as exploring outer space and caring for humans. Then there will be ever-smaller components for new sensors and actuators, with new materials. And they will have a more friendly interface and better people skills.
Are there any examples you could give of robotics projects that are making progress in these areas?
It's hard to tell. Many robots are making progress little by little.
A number of people have told us that the real

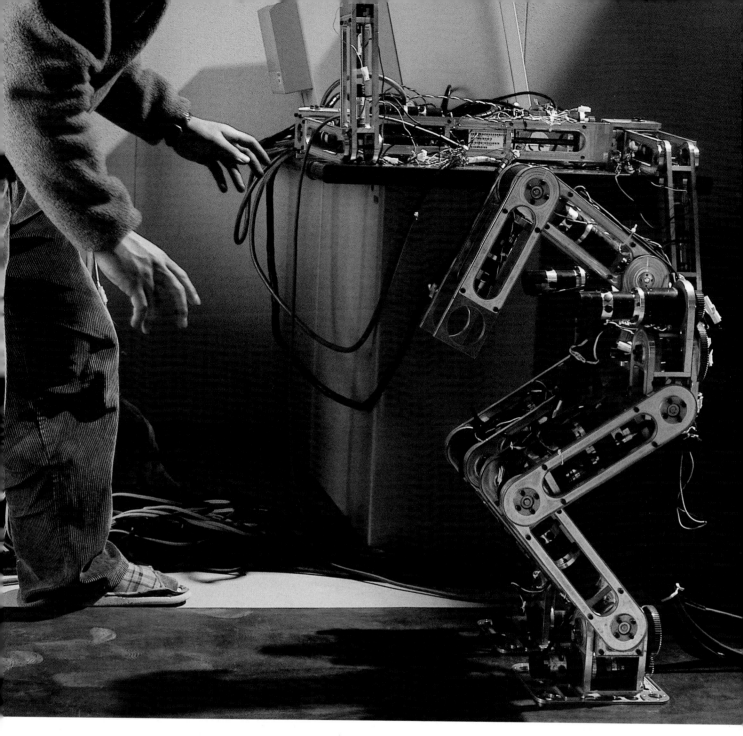

progress ultimately will come from making robots for entertainment. If entertainment robots aren't adaptable to human beings they won't be successful, so there's real economic pressure there.

It's nice to have entertainment robots like the ones from Disney and the Japanese-made robots. People need to see them occasionally, not every day—otherwise we'll get bored with them. They need more psychological computation power to react to humans appropriately. At least they must have the ability to differentiate Person A from Person B—at the moment they react to everyone in the same way.

What will people be using robots for in twenty years?

We'll probably use personal robots at home. They'll be in different shapes and structures, taking care of people who live alone. In this case, an intelligent network system with sensors and actuators spread throughout the home would be considered a single robotic system, switching on and off depending on the situation. The robot could decide to communicate with the outside world, in case of emergencies or sickness.

Okay. How about fifty years?

I don't care, since I won't be in the world then. Anyhow, by then a robotic network system will be all over the world.

Peter's Notes

On the way to the Fukuda Lab, we visited Yoji Yamada at the Toyota Technical Institute. Because workers have been killed by industrial robots, Yamada works with robots that create and measure pain. The intent—less alarming than it sounds—is to develop sensors and programs that will tell an industrial robot to shut down if it hits a worker. If the machine just brushes a worker, Yamada says, why pull the plug? But if it really whacks somebody, *stop the presses*—serious injury or death is not acceptable.

Strutting Its Stuff

Bot Specs

Name: Strut

Origin of Name: We hope the robot walks proudly.

Creative Inspiration: 1) Myself whose motion is actuated and controlled by muscle and nervous system and 2) the skillful motion of animals such as horses, camels, and kangaroos.

Height: 1.26m

Weight: 37.5kg

Vision: None

Sensors: 12 foot pressure sensors, 14 rotary encoders, 2 inclinometers

Frame composition: Aluminum alloy. The feet are CFRP (Carbon Fiber Reinforced Plastic).

Actuators: 14 electric motors, DC servomotor

Batteries: None, power supplied by cables. DC external power supply.

KLOC: 3

Cost: $80,000

Project status: Ongoing until summer 2000, then new biped robot project will start. Strut is 6th generation.

Information from: Junji Furusho

Peter's Notes

Near our downtown Osaka hotel, the neon-lit streets along the Dotombori-gawa River are lined with gaudy restaurants that include giant animated crab signs, a clown-robot playing a drum, and girls in big-soled shoes storking through the rush-hour crowds. Furusho's labs, almost as crowded as the streets, smell of machine oil and student labor, but I am slightly disappointed to find them housed in a typical Japanese university building (sparsely utilitarian, un-Zenlike) on a sprawling campus on the edge of the city.

Posing for a portrait at the Osaka University Department of Computer-Controlled Mechanical Systems, Junji Furusho (far right, seated) and research associate Masamichi Sakaguchi show off Strut, their child-sized humanoid robot. At the time, the robot (right), a work in progress, could not walk at all—it could only stand. (It walked sometime later.) But simply getting the robot to stand properly was a major accomplishment. Like a human being, Strut has such complex, interreacting mechanical "musculature" that considerable processing power is needed simply to keep it erect.

The coltish "scamper" robots that delighted us on a videotape in Professor Junji Furusho's office are disappointingly still in the corner of his mechanical engineering lab at Osaka University. "They're old," Furusho says, dismissively. He wants to show us his new biped robot. Every university robotics lab in Japan is working on one, it seems. Seeing our approach, a student in Furusho's lab hastily pulls two bright-colored stuffed animals off a robotic arm and shoves them into a drawer. "Mascots," he offers sheepishly.

Although the university is closed—it's National Culture Day in Japan—Furusho's students are hard at work on a new biped robot. It's short in stature, and has thick platform feet and a faceless head that students sometimes dress up with a demon mask, which hangs nearby.

The robot's only accomplishment to date has been to stand up—a milestone of more significance than may be apparent to the casual observer. To stand erect, humans need to send energy to their leg muscles, constantly balancing the tension in them. Similarly, robots must constantly send power to their legs to stay upright. If that energy or power was cut off, both people and robots would collapse in a heap like a rag doll. "The robot," Furusho says, "is named Strut because we hope it will walk proudly."

Faith: Why do you want to build robots that look like people? It's obviously a difficult task.
Junji Furusho: I'd like to build mechanical creatures that are like human beings. There are many control systems in a human being—for example, arm muscles have biological sensors and actuators and systems that control them. This interests me. *But engineering such a thing mechanically is incredibly complicated. Everywhere we go in Japan, we see people who have been struggling for years to do it.* Human beings and animals are very sophisticated, so I can only replicate a little part of what they do. Animals and humans have muscles, and spring mechanisms in their tendons, and they use these systems skillfully to save a lot of energy.

We can work for a whole day and eat only a little food, but the Honda robot [see page 43] works only 15 minutes before using up a large battery. So human beings have a very good energy conservation system—mechanically, they're very energy efficient.
What's your goal in building robots like Strut?
Our studies will be useful for the care of paralyzed patients. If we can understand how to control the mechanism for walking, we can help people.
But it's not only research, right? You want Strut's descendents to be working among people.
I think we won't be able to build a real humanoid robot until maybe a hundred years from now, or more. When a human needs to steady himself, he can reach out to grab something stable. But if a robot reaches out to steady itself on you, it will kill you. The Honda robot cannot judge whether something is a human being or an object.
No humanoid robots for a hundred years? How about fifty years?
Humanoid robots will not yet have come into existence. But a very simple robot—one to aid the handicapped or elderly, to feed them or for rehabilitation—such systems I suppose will be around. But a robotic equivalent of a human being is a very great challenge.

DB, the Dynamic Brain

Robo Specs

Name: DB

Origin of name: Dynamic Brain

Purpose: Experimental tool to explore issues of biological motor control, theories of neuroscience, machine learning, and general-purpose biomimetic robotics.

Height: 190 cm

Weight: 90 kg

Vision: Two 2-degree-of-freedom eyes (pan and tilt), each with two cameras (foveal and wide angle, corresponding to human foveal and peripheral vision)

Sensors: Four cameras, an artificial vestibular system (3-axis angular velocity, 3-axis translational acceleration), load sensors in 26 degrees of freedom, position sensors in all degrees of freedom

Frame composition: Lightweight aluminum

Batteries: None

KLOC: 100

Cost: $1 million

Project status: Ongoing

Information from: Stefan Schaal

Peter's Notes

After a 36-minute train ride from Kyoto, we struggle through the elevated Takanohara Station with all my equipment before surrendering $20 to a white-gloved taxi driver for a five-minute ride to the Advanced Telecommunications Research Institute Lab. ATR is international Japanese: we get name tags from a receptionist, but don't have to remove our shoes and wear company slippers. The DB project's lab is very spacious by Japanese standards and—convenient for me—it has built-in overhead spotlights for videotaping the robot. DB itself, sheathed in acrylic armor, looks a little like a technowarrior, or would if it weren't bolted to a stand, with its legs dangling. In the corner next to DB is a disarmingly simple-looking one-armed machine. Stefan Schaal warns, "Be careful of the arm—it's a real robot that can rip your guts out. It can move very quick, in 200 milliseconds—there is no way a human being can get out of the way." DB, by contrast, is better behaved. He could bruise us but isn't capable of deadly force. One small step for the robot; one slightly larger step for the safety of man.

Stefan Schaal is searching through the hard drive of a lab computer for Okinawan folk music. "I'm not a deejay," he complains in a rather teasing fashion, but despite his words he quickly finds what he's looking for: traditional Japanese dance music. The sound of a Japanese musical instrument—the *samisen*—fills the room, and a few feet away, a $1 million robot begins to dance.

This robot is the newest research platform of the Kawato Dynamic Brain Project at the Japanese research center ATR [Advanced Telecommunications Research Institute], outside Kyoto.

Neither project director Mitsuo Kawato nor Schaal, who is primarily based at the University of Southern California, are roboticists. Instead, they are neurophysicists, who are using robots to study human brain function under a five-year grant from the Japanese government. Their first project was a robotic arm that reproduced the flexibility and dexterity of a human arm—Schaal calls this robot his "workhorse." The second project is DB—Dynamic Brain. Schaal designed its specifications and developed its core programs with researcher Tomahiro Shibata, but the robot was built by SARCOS, a sophisticated robotics company in Utah.

"In order to understand the human brain, we need to understand our body and its interaction with the external world," Kawato says. "I think it's important to have a humanlike creature to study computational neuroscience. You'd like to have a good simulation of the human body, but a good simulation is difficult because we don't know how to make artificial muscles, artificial skin."

"This robot's body is one of the closest to a human body, in the sense that its height and weight are close. It's really different from other types of humanoid robots. Usually, robots are very stiff and heavy, so they are very dangerous. But you can physically interact with this robot because it's compliant and light." "It won't haul off and whack you?" I ask. "It can move very quickly, but still it's not so dangerous," he says, laughing.

Kawato organized his team of researchers into three very different groups: one that examines brain activity with a technique known as functional magnetic resonance imaging (fMRI) which tracks blood flow in neural tissue; a second that builds physiological models of the brain based on data from monkeys; and a third that builds robots in an attempt to understand the interactions of brain and body. The elegant DB is a product of the third group.

Earlier, Kawato has shown us a video of a woman dancing to this same music. Now the robot is swaying and moving its arms in a dance that vaguely resembles the one on the video. The robot doesn't dance well, but we are amazed to see that it can at all. Most surprising is how the robot learned the dance—not by following the dictates of a computer program, but by watching the video and imitating the woman's movements.

Faith: Why did you decide to teach a robot to dance?
Mitsuo Kawato: We are interested in imitation because we believe that somehow the essence of human intelligence is imitation. We learn various kinds of tasks by watching, so we recorded human movement for DB to learn from. The traditional Okinawan folk dance was implemented onto DB, but because DB's degrees of freedom [its movement abilities] are different from those of humans, it couldn't physically perform all the movements in the dance, so it was a difficult program.

One of the most striking achievements in this area was done by Christopher Atkeson [of Georgia Tech], open-loop juggling. I can't juggle. So it's nice to see a robot doing it. [Open-loop juggling is the technical term for juggling without feedback, blindfolded juggling in this case, as the robot did not use its vision, just its repeat movement, to not lose the balls.]
Is it difficult, though, seeing a robot doing something that you can't do?
Kawato: Oh, robots can already do many things that I can't. For example, the dexterous arm robot can balance a pole in the palm of its hand for two or three hours. For me, 30 seconds is my longest achievement. One minute, forget it!
Stefan, how does teaching a robot how to balance a pole in the palm of its hand teach you something about the human brain?
Stefan Schaal: We are looking into principles of learning and self-organization that allow a system to develop and become intelligent by itself. A major premise in biology is that there is structure in the world and this structure is recognized by the brain. Based on this structure, the brain can bootstrap itself, through learning and self-organization, to become better and better.

This is the process that we would like to understand. It's clear this can't happen from nothing, so what must be happening is that the genome is giving us some information about how to learn—what the right learning algorithms are for the problems that the system is facing. And we have to figure out what the basic ideas or these biases are which come from the genetic coding. But from there onward we believe that the brain can self-organize.

This is a different way of thinking. It is basically, in the end, a huge neural network that self-organizes automatically to reach this level of intelligence and competence that we see in biological systems.

In the textbooks, a neural network is a network of many simple units with simple capacities. These

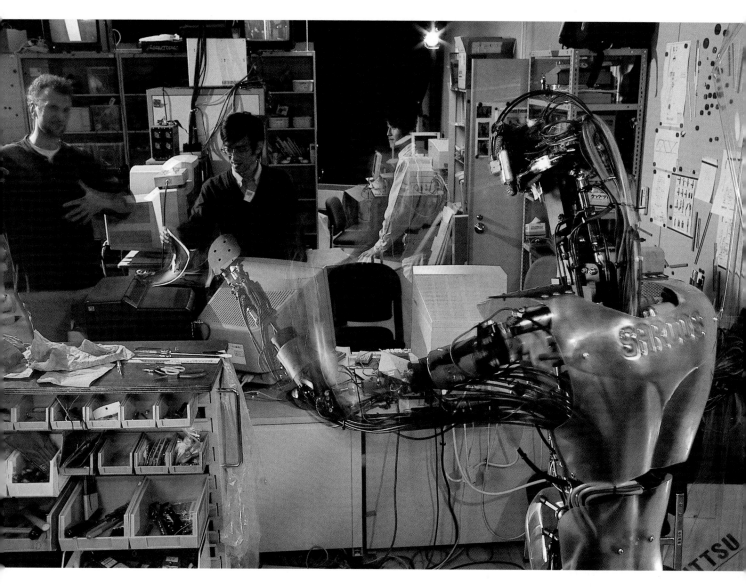

units are connected in such a way that the connections can change depending on the circumstances. The idea, which became widely known in the 1980s, is to model the neurons of the brain, which can knit themselves into new configurations to create memories and skills.

But when I ask Schaal about this, he says, "We aren't doing neural networking in the old-fashioned sense. This was pretty much gone at the end of the eighties. It's much more statistical learning—we have a very clear understanding about the statical properties of the learning systems we use." In statistical learning, the machine has a statistical model of how its sensory data are generated, and it uses learning algorithms that exploit this knowledge to assure statistical convergence to good learning results. DB learns much more from "statistical insights," Schaal says, than from any attempt by the researcher to say, "Oh, this is how the brain seems to connect neurons, let's put it together this way and try to understand it later."

Faith: I'm puzzled—I don't see the connection between what they're doing and the human brain.

Stefan Schaal: From a biological point of view, what we are inspired by is the cerebellum. The cerebellum is a huge learning machine, a real-time learning machine. It has tremendous capacity, and we don't know how to duplicate that. What we have developed is not getting even close to the cerebellum, but I think it outperforms anything out there, in this domain.

I would never say that we have something that is better than everybody else. But in this domain, when it comes to motor learning, nobody can compete with what we are developing. But we have been developing DB especially for this purpose. So it is, like, we have a special domain and there we are good. But if you put DB in another domain, it will suck.

You know your limitations.

Everything is limited! There is nothing that is not. We know that there is not going to be an omnipotent neural network or theory. Everything can only work in a certain domain. And all you have to do is find your domain and understand how to put domain knowledge into the system.

Exemplifying the attempts by Japanese researchers to put a friendly face on their robots, DB's creators are teaching it the Kachashi, an Okinawan folk dance. Unlike most robots, DB (above) did not acquire the dance by being programmed. Instead, it observed human dancers—project researchers, actually—and repeatedly attempted to mimic their behavior until it was successful. Project member Stefan Schaal, a neurophysicist at the University of Southern California (in red shirt), believes that by means of this learning process robots will ultimately develop a more flexible intelligence. It will also lead, he hopes, to a better understanding of the human brain. Naturally, the project has a long way to go. Thus far, DB has only taught its upper torso to dance; its legs have yet to make contact with the floor.

Showing off its dexterity, DB slowly juggles three small round beanbags under the alert supervision of researcher Tomohiro Shibata. The DB project is funded by the Exploratory Research for Advanced Technology (ERATO) Humanoid Project and led by independent researcher Mitsuo Kowato. Based at a research facility 30 miles outside of Kyoto, Kowato began work by adapting a robot designed by SARCOS, a Utah robotics company. At a cost of almost $1 million, SARCOS built the adapted design and shipped it to Japan, where ERATO workers continue to tinker with it. At the time, a chief frustration was DB's inability to process feedback from its own actions quickly enough—it couldn't, for example, adjust its arm motions to compensate when it made a slight mistake juggling. Although it could see the ball on its vision system, it was unable to use the information to reposition its hands. Instead DB relied on its programmers to find the perfect rhythm—which meant, in practice, that it could juggle for only a few seconds before the balls fell to the floor.

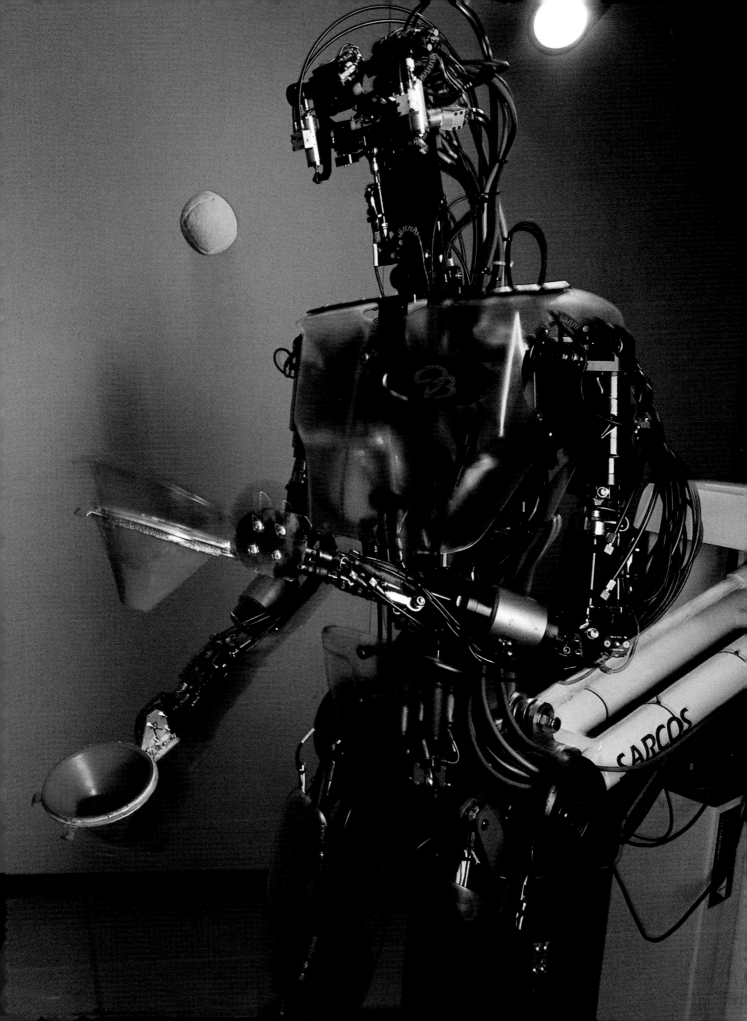

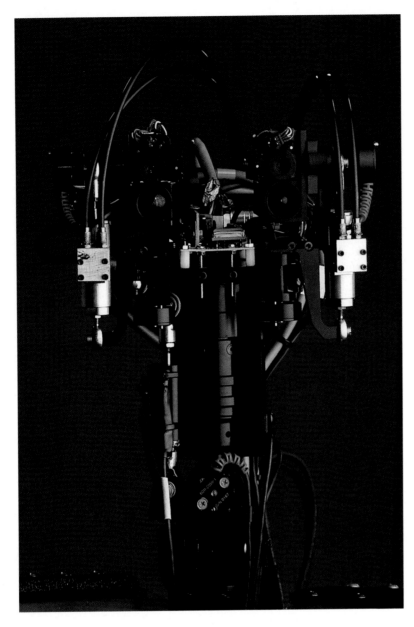

DB (above) gazes intently at the camera by means of two pairs of lenses in each "eye." In a configuration increasingly common in humanoid robots, one lens in each pair sharply focuses on the center of the visual field while the other gives a broader perspective. These two points of view, surprisingly, mimic the human eye, which seamlessly blends together information from the fovea centralis, a small area of precise focus in the center of the retina, and the parafovea, a larger, but much less acute area surrounding the fovea. By creating a simulacrum of the human eye, the DB project leader and biophysicist Mitsuo Kawato (right) hopes to learn more about human vision. Similarly, DB has a vestibular system in its ears—vestibular systems being the inner-ear mechanisms that people use to balance themselves.

How did you pick your domain?

Everybody here follows a particular branch of research where they are interested in a particular question. Tom Shibata's primary work is on the ocular motor system [vision]. And [researcher] Shin'ya Kotosaka-san is using biological pattern generators for this robot. Invertebrates and higher vertebrates are known to have kind of oscillatory pattern generators [parts of their brain that send wavelike impulses to the muscles], which is a different way of generating movement. It's primarily being studied in little animals, like cockroaches, but could also be useful for humanlike things.

We have been studying how to take these same ideas into the context of human motion generation. We're also investigating in fMRI experiments whether these oscillator theories are actually substantiated. A lot of good evidence suggests that this might be happening in the human brain and so we are just trying to mimic these theories and put them in the context of robotic work.

So the statistical method—sending out impulses and having them converge on the right behavior—isn't that different from what's happening in the brain after all?

Shin'ya Kotosaka has also been working on drumming, which is sort of the prototypical rhythmic movement. It's also an example of discrete movement—you have to go from one drum to another drum—and an example of how you synchronize with external stimuli. So what he has been working on is a robot drummer that can synchronize with external music. That is actually a very interesting research topic, from our point of view.

Who decided to use the SARCOS humanoid robot as your research platform?

There was nothing to decide. It is just unique in the world. Steve Jacobsen [roboticist and founder of SARCOS] is a unique person. He basically comes from biomechanics. He built prosthetic devices for people and then he moved into robotic entertainment and noticed that there was nothing in normal industry that could be used to build an interesting humanoid robot.

Everything was big and heavy and terrible. So he decided just to develop everything himself, and he developed his own hydraulic valves, his own sensors, his own electronic boards—everything that is important to make these systems the way they are, and there is just no competition in the world. Nobody else can do that.

What do you think of humanoid projects like Cog at MIT [see page 58], where instead of outsourcing to a company like SARCOS they built their own robot?

It is a great project that they're attempting at MIT. The problem is that they're trying to build everything themselves, which is, from an engineering point of view, a terrible thing to do. A robot is not something that you want to have work for one day or one or two weeks. You want it to work for a couple of years. SARCOS does proper engineering and is only interested in the engineering part. They don't give a damn about the controlling part—they just build enormously good hardware. And it lasts.

We have had the robotic arm, which was also built by SARCOS, for five years. We have really badly abused it and it has had no problems surviving. Since DB is very new, it has a little wear and tear at certain points, but this is because it is a new machine. It is very easy to replace these things.

Still, there are some similarities between your project and the Cog project.

They have the goal of making something cognitive, a robot that can think with self-awareness, like a human being. We are, I think, maybe a little bit more boring. We just basically want to have a humanoid robot to do computational neuroscience. And we focus on very boring details of robotics. We focus on very detailed small problems. We make very incremental progress with these things.

We don't have a point of view about cognition or consciousness, or believe that the robot is ever necessarily going to be conscious. We believe, as Kawato says, that this might happen by itself at some point, if we just do our work and understand more and more about the brain.

Right now I think it is not necessarily the time to ask this question about consciousness or to ask about how this can be implemented at some point and what it will look like. We have the feeling that consciousness is just a part of normal information processing and just requires some very high-level, complex network structure. Otherwise we just do little things—vestibulo-ocular reflex [eye movement compensating for movement of the head]; we do arm movements; reaching movements; we learn internal models.

We do what we believe happens in a real brain. We try to validate models. Get new ideas. Figure which questions to ask next, which questions to ask of biology, and continue this iteration through technology and biology.

You divide your time between USC in the United States and ATR in Japan. Is splitting your time in this way a good way to do research?

Well, it is a complicated question. If you are going to be an experimentalist, the right thing is to go away from the experiments and think for a while before you do anything. Otherwise, you just experiment all the time and you don't get anywhere.

These days, since I have pretty much the same setup at USC, I can do a huge amount of work over there and then come here with preprocessed ideas, which we can then try on these different physical systems, which are more complicated. So this is perfect. From that point of view, being here every year for two or three months is sufficient.

In the old days it was a little bit on the low side, since basically during the time when I wasn't here, nobody worked with the machines, which was not so good. We have the perfect setup between this lab and our university and now many people all of a sudden work with the robots. This is very new—we didn't have this before.

Is there a safety protocol you follow when you use the robots?

Since the robotic arm is so dangerous, we have extremely high safety regulations about who can

work with it. You couldn't just go and program the machine. It's just too dangerous. You would ruin the machine, you would ruin yourself. DB is much softer, not as heavy. If it hits you, it will not kill you, although it might make a scratch. So that is very helpful, from our point of view.

Do you have an ultimate final goal or is there just so much to do that smaller goals will suffice?

I would like to understand how the brain does perceptual motor control. The expression of that will be that we will have a machine that can do the same things people can do. But I am really fascinated by how biology works and how this entire system is set up and how it can be so incredibly competent and why we are so incredibly stupid about understanding it [laughs].

Peter's Notes

Kawato warned me before we met that in the middle of our visit there would be an hour-long, mandatory, company-wide earthquake drill. Everyone would have to flee the building. Sure enough, alarms erupted just as Faith's interview was beginning to mine the motherlode. We all looked at each other glumly. Then Kawato's secretary, on her way out of the office, handed her boss an official document exempting us from the drill. I immediately offered her a job in my studio. Later, during the photography session, I guessed that Kawato would be amenable to a conceptual portrait made by waving a ring of gold Christmas lights in front of the lens during a time exposure. Skeptical at first, he relaxed after I showed him the test Polaroid with the light squiggles. "Neuron synapses," I explained.

Face-Off

Robo Specs

Name: Now informally called Jack but researchers are still thinking/arguing. Tentatively: ETL-Humanoid.

Origin of name: Robin Williams's character in the movie *Jack* (5-year-old boy's brain in the body of 40-year-old man)

Hardware Designer: Akihiko Nagakubo

Eye Designer: Fuminori Saito

Purpose: For the study of continuous complex human/humanoid interactions, and the derivation of software/hardware mechanisms required from a globally integrated view.

Height: Approximately 1.6 m

Weight: Approximately 60 kg

Vision: Stereo vision, CCD video cameras

Sensors: Binaural auditory microphone system, joint receptors (proprioception)

Frame composition: Mostly aluminum

Batteries: Not yet determined

Cost: Custom-built components cost a lot. Overall cost (including integration) is way over $1 million.

Project status: Ongoing

Information from: Yasuo Kuniyoshi

Looking into the eyes of Jack the robot, Gordon Cheng (left) tests its response to the touch of his hand. Researchers at the Electrotechnical Lab at Tsukuba, an hour away from Tokyo, are part of a project funded by the Japan Science and Technology Agency to develop a humanoid robot as a research vehicle into complex human interactions. With the nation's population rapidly aging, the Japanese government is increasingly funding efforts to create robots that will help the elderly. Project leader Yasuo Kuniyoshi wants to create robots that are friendly and quite literally soft—the machinery will be sheathed in thick padding. In contrast to a more traditional approach, Kuniyoshi wants to program his robot to make it learn by analyzing and fully exploiting its natural constraints—unlike Honda's humanoid robot in which every movement is completely planned, and rigid control forces the system to follow the commanded trajectory. The robot's lower torso, seated nearby with legs crossed, seems to be waiting patiently for this day.

G ordon Cheng is trying to play an impromptu game of Imitate the Human with his robot, Jack, but he can't get its attention—something else has caught the its eye. "I think it's locked on you," Cheng says, eyeing me warily. I move myself out of Jack's line of sight. The robot immediately focuses on Cheng. When the researcher moves his arms, the robot moves its arms in the same way, mimicking him. "It has to lock on to me and be able to see my upper body to do this," says Cheng, who designed the programming and interface for the robot's sophisticated vision system in this lab at Tsukuba, the large science center outside Tokyo. Monitors show what the robot is seeing—Cheng's face looms large at center screen. "I like looking in his eyes to see if he's looking at me," Cheng says, staring intently at the robot. He extends his right arm outward; the robot does the same.

Faith: What happens if you turn away and then re-engage it? Does it have to start over or can it pick up where you left off?

Gordon Cheng: I can stop, talk to you, and go back, and the experiment basically continues. It's all multisensory—it has hearing and seeing. I can go in, fiddle with the robot, and it won't hurt me. It's fully compliant. It's fully backdriveable and beautifully designed.

It sounds like you're trying to sell me a car [laughter]. Do you feel like the robot is responding to you?
Sometimes. It's like working with a child and that's interesting.

If Jack is a child, it's an extremely odd one. At the moment, its lower torso is seated in a chair, legs crossed, as if waiting for something to happen; the upper torso, gyrating nearby with Cheng, doesn't have hands yet. But the machine's incomplete state has no visible effect on either Gordon's imitation game or his enthusiasm. Nor does it disturb project director Yasuo Kuniyoshi's optimism about the future. "This year," he says, "we'll show some fancy stuff."

Faith: Dr. Kuniyoshi, your work centers on the idea that a humanoid robot doesn't need to be programmed explicitly from moment to moment.
Yasuo Kuniyoshi: Our emphasis is that the robot is not optimized to a particular task, like walking, standing or grabbing. We're aiming at more general motion, like rolling around, leaning, or lifting itself [from a prone position].
Such as a human who is lying down and wants to stand does?
Yes. One way is to swing your feet and then roll for-ward. Your hands stop you, and you stand up. You use your own weight and inertia to put yourself through a trajectory to reach the goal. When you do that, you don't have to precisely control every point where you are.

So you're not programming the robot through every point, which you'd normally have to do?
Right. The critical idea is to utilize what Dr. Nagakubo and I call global dynamics. It is pointless to very accurately model or measure every point. The key to success is finding out just the right points. This is strongly related to other issues in cognitive science—focus of attention issues. The same problem structure holds

when you observe the world and try to recognize things. You are not seeing everything; you are only selecting partial aspects and jumping around among them. And you form your conceptions by integrating these selected fragments.

The Honda robot is a walking robot like yours, but it differs substantially from the robot you are building.

There are a lot of ideas milling around out in the robot world about proper walking. The Honda humanoid is very typical, very nicely done engineering work. It cannot do much else, other than walking. *It can also kick a soccer ball and open a door.*

It is basically designed for walking. The upper part assumes that it's always standing and doing a very clearly defined set of tasks—grasping, manipulating. Our humanoid is different because we assume a general posture, like lying on the floor, rolling around, leaning on the wall, or rising up with the hands and supporting itself, so that the arms must support the robot's own weight, and the legs should do much more than walk. We haven't designed our robot to do very precise manipulation tasks. And it can't do them, but that doesn't matter to us. We are interested in much more global things. We don't specify this task or that task.

COGnition

Robo Specs

Name: Cog

Origin of name: The name was chosen both to evoke the word "cogn-tive" and to refer to the mechanical teeth (cogs) on the edge of a wheel.

Purpose: To investigate themes of development, physical embodiment, sensory-motor integration, and social interaction. To study models of human intelligence by constructing them on a physical robot.

Height: 1.72 m (including 86 cm pedestal); with arms extended, robot is about 2 m across

Weight: The robot is bolted to a (very heavy) pedestal ... we have never weighed the entire assembly.

Vision: 4 color cameras (2 for foveal vision, 2 for peripheral)

Sensors: We use a variety of sensors—2 microphones for ears, a 3-axis inertial package for vestibular function, and an array of tactile sensors. We also have kinesthetic sensing from each of the joints, which include position readings (from encoders and potentiometers), force sensing (strain gauges), limit switches, and temperature sensors on the motors.

Batteries: None

External power: +/- 12 V, +24 V, and +/- 5 V

KLOC: Impossible to estimate.

Cost: Impossible to estimate.

Project status: Ongoing

Information from: Brian Scassellati

With its carapace not yet built, the mechanism inside the head of Cog (right) is revealed against a photographer's lights. Cog's designer is Rodney Brooks, head of MIT's Artificial Intelligence Laboratory, in Cambridge, Mass. Marrying the disciplines of biology, mathematics, philosophy, sociology, computer science, and robotics, Brooks has created in Cog one of the most sophisticated robots to date—a machine that he hopes will someday possess the intelligence of a six-month-old baby. Although some might be discouraged by the disparity between the enormous amount of thought and labor that went into it and the apparently meager results, Brooks draws a different conclusion. That so much is required to come close to simulating a baby's mind, he believes, only shows the fantastic complexity inherent in the task of producing an artificially intelligent humanoid robot.

Like most communities of researchers, the robotics community is a gossipy village. Other scientists regard its few stars with intense fascination. We experience this first hand, at Osaka University in Japan, while talking with students about the different robots we've seen. When I mention the robots at the Massachusetts Institute of Technology, the students grow quiet. "Rod Brooks is at MIT," a student whispers. The tone is reverential—unmistakably so.

On the other side of the world, when we visit Brooks, who is director of the storied MIT Artificial Intelligence Laboratory, the researcher is giving graduate student Brian Scassellati a critique of their humanoid robot Cog's new head. "It looks like an insect," complains Brooks, who is eye-to-eye with the robot, getting ready for Peter's photograph.

"It isn't finished yet," retorts Scassellati. "Your other robots looked like insects, too," I point out to Brooks. Actually, his earlier robots *were* insectoid.

Brooks is perhaps best known for developing the concept of "subsumption architecture" in the mid-1980s. His idea was to split the behavior of a system into simpler behaviors controlled by subunits—a departure from the conventional approach in which a central controller, or processor, makes decisions for the entire robot.

For example, in traditional robotics, the robot instructs its knees and ankles how to bend when it walks. By contrast, a robot with subsumption architecture has sensors and small computers in its knees and ankles that tell the joints how to move—the central brain is not involved. Using this concept, Brooks and his students built insect robots that could react simply and automatically to their physical environment.

Thanks to his camera-friendly bugs which could skitter around the room without having to wait for a central processor to file a flight plan, Brooks became a media darling. He also became known as something of a renegade—a reputation he solidified in a 1986 scorched-earth assessment of artificial-intelligence (AI) research which dismissed much of the work in the field. Subsumptive architecture traveled as far away as Mars when NASA incorporated the concept in the planetary rover Sojourner, on the Mars Pathfinder mission in 1996–97.

Most recently, though, Brooks has irked traditionalists with his high-profile contention that the key to deciphering human intelligence is to build a machine that experiences the world from infancy, like a human being, rather than build one that has its brain filled with ready-made experiences. Cog is the humanoid robot platform for this research. Cobbled together over a succession of iterations of ever newer technology, the robot is currently undergoing a hardware overhaul. And a dissection of long-term goals as well. I talked to both Brooks and Scassellati:

Faith: Rod, why do you think you give some AI researchers such acid stomach?
Rod Brooks: I think it upsets people that I really believe what I say. What if I am right? Then a lot of what they're doing is useless. That's threatening to people.
So it's fear?
That's my rationalization. Of course I may just be a grumpy old asshole, is the other possibility [laughs].
Are you a grumpy old asshole?
Sometimes [laughter]. It happens when you get older, you know. I developed this behavior-based subsumption approach to robots and it certainly has had an influence on the field of AI. There has been a grudging acceptance of it, but it's not the acceptance I wanted, and that's not just purely from an egotistical point of view. I really think that this is all there is in animals—and people.
This lower-level processing?
What's happened is, AI has sort of accepted this: "Oh, okay, that's how the low-level parts work, but then there must be this other high-level stuff." But I don't think there is that high-level stuff. I think that's pretty much all there is and that we overanthropomorphize people too much.
How can you overanthropomorphize people? Anthropomorphism is about attributing human characteristics to nonhuman things, but it seems pretty safe to attribute human characteristics to humans.
We attribute too much to what people are doing. I think that's part of a more general category error that people make, where they mistake the appearance of something and a description of the appearance for a description of the mechanism. So yes, we can talk about something rationally, in some logical fashion, after the fact and maybe even before the fact, but just because we can do that doesn't mean that this logical chain of reasoning is the process we use to control our behavior, most of the time. I think that the evidence suggests it isn't—certainly in my case [laughter].

You know, we are a collection of these primitive pieces that are put together, and we have built and modulated the world so that they work together, and our society has built itself, and we have all this mechanism around it, but it's largely this scaffolding that we have built up out of these primitive pieces. And I don't think that you have to postulate too much more to get us. We are like bees. The individual bee is not too good, but the bee in conjunction with the others is.

People accommodated me because I was waving the flag, but they managed to mentally put me in a box, saying, "well, that's just the low-level intelligence! Yeah, he was right, and it was good that we

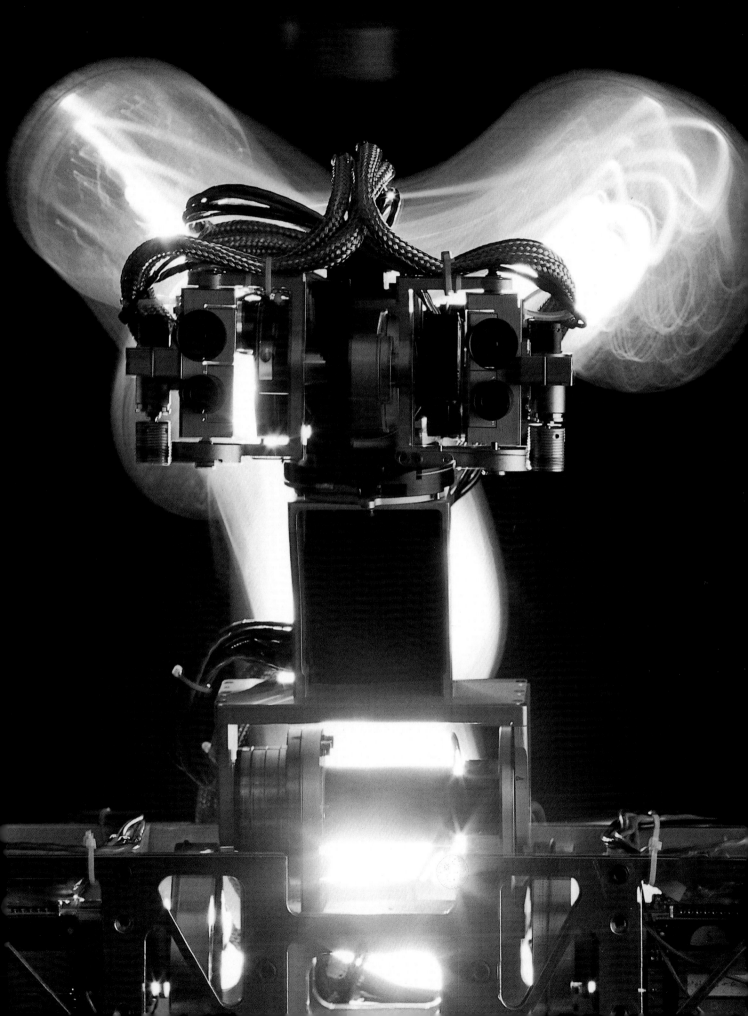

thought about it, but there is really this other level too." Well, they're off thinking about this other level, which I think is just an illusion.

What are you going to do about it?

One of my goals in life is to prove them all wrong by actually making my low level approach wildly successful.

How?

Two ways. One, I am trying to push it through with the humanoid-interaction research. And the other is at iRobot—[Brooks's robotics company that he started in 1990 with his former graduate students Helen Greiner and Colin Angle]. I'm trying to push it into real applications, which we're starting to do on a few different fronts. My goal is to fill the world up with real robots doing real applications that will, undeniably, be real applications, and will work on these principles. But then everyone will find another excuse to write me off.

You're saying there are people out there trying to do something that can't be done?

That always happens in science, but I may be totally mistaken.

You don't think you are, otherwise you wouldn't be doing it.

Of course I don't think that I am; I have sort of bet my life on it. But yeah, I could be wrong. The question is, whether, if I am wrong, will I be smart enough to see it and big enough to admit it?

You are still using the concept of subsumption architecture with the robots at your robotics company, but your focus at the AI lab has been on human intelligence for the last several years. Tell me what you think Cog is for.

I think Cog is a way of exploring what makes humans. And with Cog we're trying to build a human. We're not trying to build a functional robot to do any particular task. We're not trying to build an entertainment robot. If we were doing that, we would be going about it in a different way. So it's not an engineering object, trying to solve some engineering problem. It's a curiosity-based science and it's not quite traditional science, because it's this murky thing that people don't have an understanding of.

We're really exploring, to try to get an understanding of that murkiness. And the surprising thing—and we didn't expect this at the start—is the way people have reacted to it one-on-one, when they're with it. The human form really makes people react in certain ways.

So they're reacting to the robot's anthropomorphic embodiment, not to what it's doing?

It doesn't have to be doing much to trigger this reaction. We haven't tested this but it seems like it doesn't take much to trigger in people's brains the responses they have to people.

But as you were saying earlier, people anthropomorphize everything. They anthropomorphize chairs.

Yeah, well okay. So we're dumb [laughter]. We didn't expect it. It was a surprise to us.

I think that once you stick a face on something—

The original Cog wasn't really even intended to have a face. But even the motion of the eyes was enough of a cue. And when the eyes moved and then the head followed, that really got people.

Where is Cog in its evolutionary process?

We're upgrading its head, for a whole bunch of reasons. One, we've gotten better at building heads. The other is it's got a more humanlike anatomy now, so when it nods, it looks more humanlike. We find that people do interpret things well, if we make it easier for them to interpret, and we can get more interaction. The other thing we're doing is an upgrade on the computers. Technology has changed so many times that we had a patchwork—a reptilian brain inside—and we threw all that out.

Cog is a collaborative project between Brooks and graduate students, past and present, who have researched their way through his laboratory. Brian Scassellati is currently responsible for most of the physical work on the robot though as an undergraduate in cognitive science at MIT, Scassellati never once stepped onto the floor of a robotics lab. Now, in the PhD program, he jokes, he rarely steps off it. Although his work with Cog centers on building its social skills, he spends a considerable portion of his time with a screwdriver in his hand.

Faith: Is it difficult to determine why something isn't working when you are working on the physical robot?

Brian Scassellati: Some things are obvious. When we built the last version of the head, we tried to have the robot imitate a person's head movement. We've got great video where I stand in front of the robot and I nod my head yes and the robot nods its head yes, and I shake my head no and the robot shakes its head no. Nice demo—but there's something wrong with it. And it turns out that it looks

Although MIT roboticist Rodney Brooks (above) has worked in robotics since the late 1970s, he first attracted widespread attention when he began building robot insects, in the 1980s. (He was one of the subjects of *Fast, Cheap, and Out of Control*, a documentary film.) Although he hoped the robobugs would be put to practical use exploring the moon, and even constructed the "sandbox," a simulated lunar environment (behind him in this 1991 photograph), to test them, the machines were intended primarily to demonstrate a controversial thesis he called subsumption architecture. Arguing that robot brains that directly send walking instructions to their feet and legs don't reproduce the experience of walking, which is nearly automatic, Brooks began building robots with low-level processors that handled such tasks almost without input from the central brain. One of his team's first subsumptive robots was the insectoid Attila (far left, in photo from 1991), here being worked on by graduate student Cynthia Breazeal. The other pairs of hands belong to then-undergraduate student Mike Binnard, and former graduate student Colin Angle, who is now chief executive officer of the robotics firm iRobot. After the insect robots proved successful, Brooks realized that logically he should move up the evolutionary ladder and build robot versions of more complex organisms, as he slowly approached the task of building a humanoid robot. But he also realized that this safely logical course might take decades. So, instead he decided to go for broke and see how far he could go in developing a humanoid robot: Cog.

Sharing a Slinky moment with his creation, roboticist Rodney Brooks of the MIT Artificial Intelligence Laboratory looks over his shoulder at Cog (short for cognitive), the robot he has been developing since 1993. Although Cog can mimic human behavior, here it is using haptic feedback to adjust its actions to play a Slinky duet with Brooks. Using a visual system modeled on human eyes, it can focus on objects and extend its arms toward them, correcting itself along the way. As of now, Cog is just a head, two arms, and a torso mounted on a platform, though since this photo was taken, the machine has acquired an elegant new head. Brooks is less concerned with making it mobile than with creating a system that will let the robot reliably tell the difference between static and social objects—for instance a rock and a person. In the resolution of such apparently simple distinctions, Brooks suggests, is a key to understanding at least one type of human learning.

In 1998, when Rod Brooks first showed me Cog, I asked him if he'd seen Terry Gilliam's *Brazil*—the robot was as stylishly retro as the movie. Bolted to a thick metal pedestal, Cog sat backed up against a bank of 20 computer monitors that displayed grainy versions of what the robot saw through both its peripheral and foveal video-camera eyes. It moved the way you would expect a robotic casino dealer to move: smooth, but not smooth enough that you would think it were human—or trust it. Given the stumbling movement of the other robots we'd seen, Cog was impressive. But to me it never seemed very congenial, even when it played a Slinky duet with Brooks. Maybe Cog's mechanical musculature was too intimidating; maybe its buglike head was too unsettling. It was difficult to understand why many roboticists spoke of it so fondly—or why they thought this sawed-off android was a landmark in robotics.

The next few times we saw Cog, it hadn't changed much. Same old metal, same old mettle. Then we ran into Rod Brooks at a conference in Japan. Brooks, newly remarried, had cut off his long hair. "You have to see Cog now," he told us. "He's got a totally new head, too."

Back in Cambridge we saw Cog again. Not only was the robot new and improved, but so were the computer monitors in the background, especially the huge flat-screen monitor on the wall. But it was Brian Scassellati, a behind-the-scenes researcher, who made our day. Scassellati had spent four solid years working on this machine. He was, he confessed, "like the opposite of a Maytag repairman—I spend all my days fixing the robot."

Faith kept asking him why Cog was important. Using a physical body in tests, he insisted, was fundamental to real understanding—computer models weren't enough. Before architects build a new structure, they construct a model. When they get new ideas, they build new models. Cog was a model of a social robot that was slowly and methodically evolving. Why go to all this trouble? Why build a sociable robot? The real reason, Scassellati said, was to learn something about the functioning of *Homo sapiens*. Every tiny step on the long road to building this bug-headed, wrestler-torsoed robot, in other words, was an additional bit of knowledge about what it means to be a human being.

very right when the robot moves its head no but very wrong when the robot moves its head yes. The reason is that it would swing its whole head and neck backward and forward and look like it was trying to head-butt you or something [laughter].

And as you continue working on the physical body, you're trying out models of human behavior on Cog?
Part of what I do is look for models in cognitive science that are developed enough that we can actually implement them on a platform—and Cog is that platform. I've always been interested in how children understand the world around them—how people get these enormous amounts of information and skill that they don't start out with.

The model of development that I work closest with right now is called joint reference. It's this idea that I can get you to pay attention to something without speaking or telling you anything. I can point at it, or I can look over at it, or I can orient toward it. And in all these ways I can get you to pay attention to the same thing that I'm attending to.

That's part of the research you're doing on autism?
Yes. Child psychologists are interested in it because children develop this skill over a period of two years. So it takes you two years to learn this. Even after a child begins to walk and talk, this joint reference skill and the understanding of other minds still haven't been mastered.

And it's not something that one can develop on one's own.
It needs to be a social skill. There's interest from the developmental psychology community because of autism—in which this development seems to be remarkably deficient. There's interest from the philosophy community because they want to know what it is to be a mind and understand another mind and what implications that has.

You're interested in how the robot performs in these models but you are also interested in how humans relate to the robot. Give me an example.
It's strange, what people respond to. If I turn the robot and set it so that it's looking at motion, it's looking around at things that move, and I just have the eyes moving, people are interested in it—kind of curious—but it doesn't quite grab them. If I set it so that the eyes turn to look at an object and the head turns and orients toward them, it's suddenly very hard to ignore.

It does feel vaguely unnerving—I'm sure most people tell you that. I almost expect it to get up and come over, but it can't. It has no legs.
I can't leave the robot running while I'm sitting there typing because it distracts me every time [laughs]. I know exactly what's happening with the robot, I know exactly what it's responding to—and it doesn't matter.

Isn't that the reaction you want?
I really want that feeling that there's another entity

here—the reality that there's another thing here that has goals and maybe beliefs and desires and intentions. That's important. That's critically important, at least to this model of autism that we're looking at. But I also think that its critical to everything else we understand.

It's funny when we bring in a tour of schoolchildren and let them play with the robot. When they first come in, usually the robot is not moving. They come up and poke at it and look at it. It'll be interesting, but not all that fun. But when the robot turns its head and looks at them, all of a sudden it's a completely different kind of toy, because it's responsive.

It changes the whole way they treat this piece of metal. And in some ways that's so very important and so built into our understanding of what the world is like—that there are things out there that are social, and there are things out there that are not social. And how you can get a robot to respond, the ways these things react, that's really what I'm interested in.

The contrast between Brian Scassellati and Rodney Brooks seems obvious—ambitious young researcher working all night with a screwdriver vs. the grand old man of the field, less inclined to be hands-on but taking the longer view. So obvious, in fact, that I wonder if it is actually true. The only way to find out, of course, is to ask Brooks himself.

Faith: Do you physically build anything yourself anymore? I know you did when you were a kid and when you were doing the insect robots.
Rod Brooks: I still sometimes get involved. And I certainly get involved with the wires and the programming over at ISR [now iRobot]—more so than I do here. I get involved at the physical level and build connectors and connect stuff up.

Is it important to do this?
I do it because I like it, because it is the most fun. And also, you get a feel for what is going on. If you just sit back in isolation, you have this disembodied notion of what intelligence is and you are not thinking of what the senses are really delivering. Yes, getting your hands dirty is important.

If the general public has any idea about what you're doing, it comes from the movie Fast, Cheap, and Out of Control [*the 1997 documentary film by Errol Morris*]. *We saw it twice.*
That's more than most people. My daughter fell asleep in it twice [laughter].

Don't get a big head. We had to see it again before this conversation because we couldn't sit through it the first time. You were okay but mole-rat-man made me crazy. Anyway, I digressed—
I am interested in what it is that makes us. I am interested in the nature of life—what it is, why things are living. What is the difference between

living and nonliving? It's all still pretty fuzzy. I am interested in all of those sorts of things. I have worked on pieces of these questions; I am not just trying to build robots for the robots' sake. Although, building robots sure is fun [laughter].

Do you wish that more people supported your ideas?
Yeah. On the other hand, I might feel that when they begin believing it, it's time to go do something else. You know, I was complaining to somebody a few weeks ago that it was much more fun when I was the lone young rebel, and it is nowhere near as fun to govern.

You're not an old guy—even though you talk about yourself as if you are—but you do have certain responsiblities; you're running the AI Lab and in the field you have achieved a certain level of responsibility.
I have also accepted that level of responsibility. I have been wondering whether I should, whether that was the right thing. Basically it takes up 90 percent of my time—keeping people happy. I am on sabbatical currently, and today I have been running around in negotiations on people's jobs and salaries who are not in my research group at all, just keeping things

humming. It is hard to get away and spend time to think. I like to think that I can still make some contributions, but how to achieve that balance?

Over all, though, are you comfortable in your shoes?
No, in the sense that I wish that I had been able to prove my truth to more people. Maybe I have just worn them down, rather than convinced them. We've had a behavior-based robot on Mars and another one that was tele-operated, so yeah, there were successes, but at the same time I always want more. You know, if I were going to write an autobiography, it would be called *More.*

What's the goal of More? *When does it stop?*
I would like it if the things that I do really have a real impact on the world.

What kind of impact though, the creation of human-like artificial life?
That's what I'm trying to do with Cog, but I don't expect to have payoff for a long time.

How long?
Maybe fifty years, maybe ten years. Ten years is just too optimistic. But it's going to be a long time.

Looking at Cog's still unfinished head, MIT neuroscientist Brian Scassellati (above) ponders where he should mount the microphones that will enable the robot to hear. As important as controlling how the robot responds to people, he believes, is having some control over how people respond to the robot. The project does not have the resources to create a mock human being. Nor would that necessarily be desirable, he says. But at the same time the robot should not look too unfamiliar. The trade-off is complex and uncertain—just another one of the huge number of intellectual puzzles that must be resolved before the first fully humanoid robot can mingle freely with its human creators.

Expressionist

Robo Specs

Name: Kismet

Origin of name: From the Turkish word for "fate."

Creative inspiration: *Star Wars*

Purpose: Natural and intuitive social interaction with people, socially situated learning.

Height: 38 cm

Weight: 4.5-7 kg—not including the huge computer rack.

Vision: 3 digital cameras (2 foveal, 1 peripheral)

Sensors: 3 microphones, 21 encoders

Frame composition: Aluminum

Batteries: None

External power: 6A 24V supply, 16A 12V supply, 5V supplies for electronics

Computers: Lots. (4 Motorola 68332s, 9 400Mhz PCs,1 dual-450Mhz PC, 2-500Mhz PCs)

Person-hours to develop software: Oh God, I don't even want to think about it…. There's tons of infra-structure code that isn't specifically for this robot. Code to specifically run Kismet is probably two full-time people working for 2.5 years. The total size of all the software tools we have developed to support our computation environment is huge.

KLOC: Are you kidding?

Cost: $25,000 (physical materials)

Project status: Ongoing

Information from: Cynthia Breazeal

Wedged into her small, cluttered workspace in the MIT Artificial Intelligence Laboratory in Cambridge, Mass., researcher Cynthia Breazeal (right) holds a mirror to Kismet, the robot head she has been working on for two years. The cameras behind Kismet's big blue eyes send data to its computer, which has software allowing the robot to detect people and bright toys visually. In addition, the software recognizes people by vocal affect. Then, following its pro-gramming, it reacts, twisting its features in a comically exaggerated display of emotion. Spread across a backlit surface like a Kandinsky painting, the disassembled Kismet head reveals the mechanisms (far right, an updated second-genera-tion version with a neck that "cranes") that allow it to manipu-late its cartoonish lips, eyes, and ears into expressions that seem startlingly human.

I f Kismet the robot were a child, Cynthia Breazeal would be its doting parent. A long-time student at the Massachusetts Institute of Technology, Breazeal is obviously charmed by her creation, which is intended to communicate with people, understanding their emotions and displaying its own. The diminu-tive researcher smiles just as broadly as the robot does when it reacts happily to the stuffed toy she waves in its face.

The pink-eared, rubbery-lipped Kismet alterna-tively pouts, frowns, and displays anger, along with a host of other expressions that outwardly display signs of human emotion. This robot began its exis-tence as the vision platform for Cog [see page 58], the MIT Artificial Intelligence Laboratory's humanoid robot project, but it is now the physical embodiment of Breazeal's study of human-robot interaction. Working with Rodney Brooks's AI Lab colleagues Brian Scassellati, Paul Fitzpatrick, and Lijin Aryananda, Breazeal is trying to build an artifi-cial system that learns in an intelligent and social manner.

In her laboratory, I ask how she went from the rather sober Cog project, for which she was chief architect in the 1990s, to the more whimsi-cal Kismet. In a rapid burst of the jargon favored by roboticists, Breazeal tells me that after Cog she decided that she wanted to focus on "face-to-face, caregiver-infant, socially-situated kind of learning"—the kind of learning in which a baby, guided by its parents' social and emotional signals, learns how to behave around other people. I ask the newly-minted MIT PhD why she hadn't adapted Cog for this purpose.

Cynthia Breazeal: Cog wasn't appropriate for that. First of all, it's huge. Nobody wants to treat that robot like an infant. There is a whole different kind of interaction, dynamically. It's a very intimi-dating kind of structure face to face because of its size. So I wanted something kind of cute and appealing that you would want to baby in the right way—to interact with—without having to think about it.

Faith: Why a robot with facial expressions?
As we interact with other people, we expect these social cues to be coming back to us. It helps us to understand what the other person is doing. So it's important for the robot not only to understand those cues coming from a person, but to send those same cues back.

Really, it's a dynamic between the two—you are not just looking at the robot as an isolated learning system, you are looking at the learning system of the person as well, because the person also adapts to the robot.

Are these generic social cues, or do you think your own personal social cues are coming into play? In other words, is it learning how to interact with people in general or with Cynthia Breazeal?
I'm sure that's probably happening to some extent, just because I built it. The goal, of course, is that anyone coming off the street could interact with the robot in a natural way. They won't have to think, "Oh, this is a robot, I have to modify my behavior in all these different ways." That's a long-term goal, obviously, but that's the goal of this project.

And so you're approaching this goal by working on early childhood learning experiences for your robot.
Right now, we're working with toys and with faces.

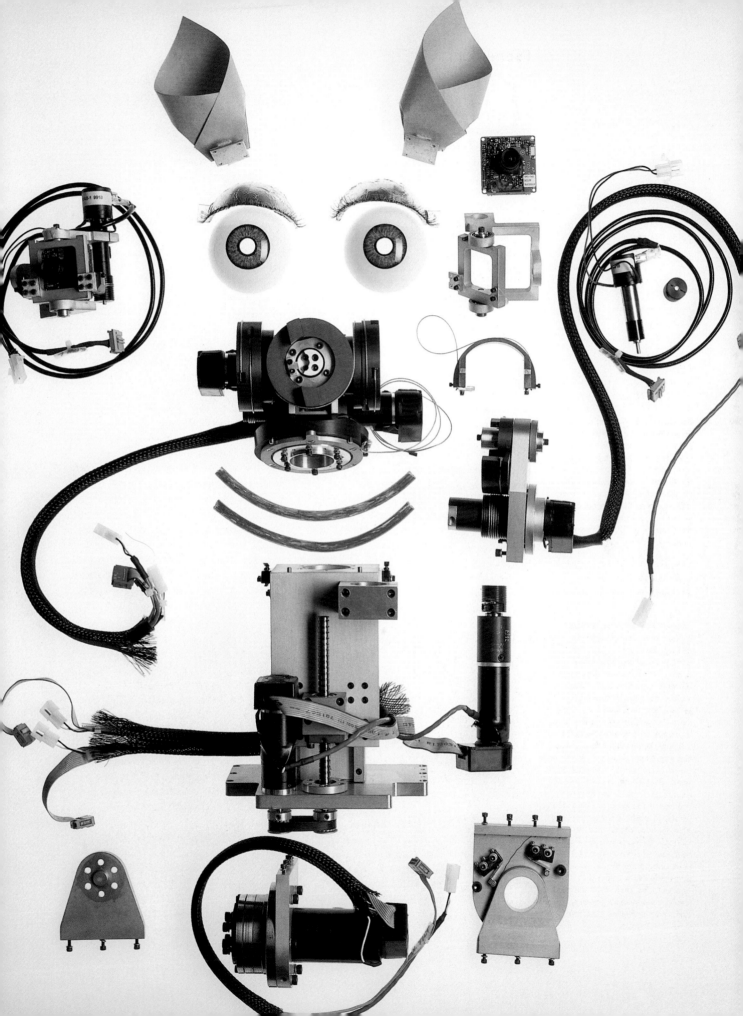

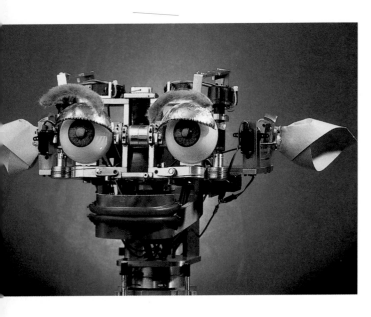

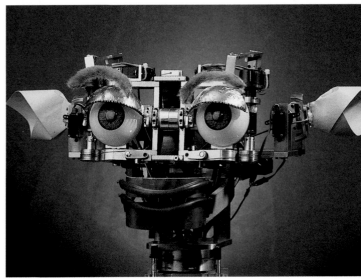

Eyes sweeping the room with what seems to be hopeful curiosity, Kismet the robot sits like an animated bust on Cynthia Breazeal's desk at MIT. When it spots visitors, the robot's expression changes to an almost uncannily convincing expression of interest and delight. (It is now able to crane its neck and shift its head from side to side as well, further enhancing its expressiveness.) What happens next depends partly, of course, on the visitors. If they wave their hands close to Kismet's face, it looks annoyed; if they show it bright colors, it smiles; if they don't do anything, it actively seeks out something else of interest—unless its fatigue drive is strong enough to induce sleep behavior. Watching its eyes lock into theirs, many visitors report being unable to stop responding to Kismet, as if it were a human baby in mechanical disguise—they find it almost impossible not to anthropomorphize the machine. (Above, Kismet showing, from left to right, calmness, happiness, sadness, and anger.)

We're starting to work on these imitation games, which constitute a very important tool children use to learn. Children mimic their parents a lot. It's a way to have them learn new ways of shaping their behavior, so they can try the things they've learned in other contexts, and learn the results of that. Like babbling freely. A child comes up with a word that has been reinforced in a different situation. Maybe the child said its name in a particular situation that was very appropriate and the parent gave it a lot of positive reinforcement. That's how it can learn to apply behaviors to new situations.

Your background is in electrical engineering and computer science, but this sounds more like child psychology than robotics.

I have a minor in brain cognitive science. This lab is unusual in that we have always taken a very heavy inspiration from biology in order to build our robots. We're the AI Lab. We want to understand all forms of intelligence, animals as well as humans, so we test and adapt theories and ideas—from seeing how natural systems work, to our robot—to see if we can get the machine to exhibit a similar kind of behavior.

Can they give us any insight into how people must be addressing similar issues and what the issues even are?

It really helps us tease apart and debug a lot of theories, in that theories tend to be very general and they leave out a lot of details because they have to— they don't know the answers yet. But we can take the broad sweep of it and try to gain insight into, "Well, here's the big picture of the theory but here are all the little issues that have to be addressed in order for this theory to hold water—and this is how we've addressed them, which may not hold true for

biology but at least we know those issues must exist." So it helps direct further experiments as far as—this is an issue? How do people deal with that? So we hope for a rich interchange between the natural sciences.

How did you design Kismet to deal with social cues?

The idea is that the robot has drives that have to be satiated. These are basic needs—like hunger and thirst for us. The robot has a need to be stimulated by people; it has a social drive; it has a need to play with toys—to be stimulated by objects—which we call stimulation drive. It also has a fatigue syndrome, so over time it gets tired.

The robot wants to keep these drives in homeostasis, a relatively stable balance—a slightly positive and slightly aroused state, which gives it an expression of optimistic curiosity toward the world.

Why do you want that homeostasis—just because it seems like a pleasant state?

In general, here's the big idea: the robot's emotions and drives are used to regulate its interactions with a person. The facial expressions that arise from these motivational states are just one form of social cue that influences the person's behavior. The robot uses these facial expressions to tune the human's behavior so that it is appropriate for the robot—not too much, not too little, just right. For example, a withdrawal response, or a look of annoyance, when the person is too close establishes a personal space for the robot.

It also helps the robot's vision by keeping people back at an appropriate distance instead of in the robot's face. If it's done well, then both the robot and the human are well matched and interacting in a way that benefits both. In addition to facial expressions, other social cues the robot uses

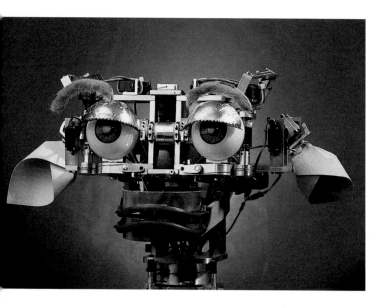
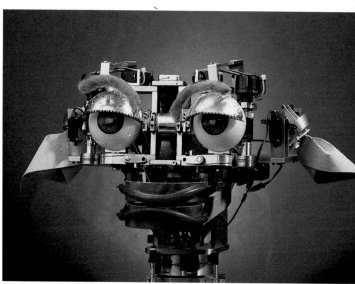

include gaze direction and mutual regard looking into a person's eyes. Gaze direction is a cue that tells the human what the robot is currently interested in. Mutual regard tells the human that the robot is directing the interaction specifically at that person. For instance, if the robot does a greeting gesture, a person will not interpret this action as addressed to them unless the robot looks at them while doing it. If the robot looks at another person, the first person will assume the robot is greeting that other person. If the robot looks at the person's neck, he or she won't interpret the robot's greeting as sincere but as just going through the motions. Kismet's direction of gaze is one of its most powerful social cues.

Other cues are gestures and facial displays, such as raising the eyebrows and tilting the head a bit, when it's awaiting a response. Kismet does all of these and more.

So by using its repertoire of behavior—facial expressions, mutual regard, and so on—it tries to get people to do what it wants?

Our idea is that the robot is always engaged in this game of trying to keep its drives in balance. The robot is specifically designed so that it needs a person to satisfy these drives—it cannot satisfy its drives on its own. It's designed that way to really force it to interact with people. The stimulation drive is very intense so it's very much in need of some sort of toy. It will look for something that is colorful—a characteristic of toys. If it finds something like a colorful block, it will look at it with a look of happiness because it has found what it wanted. If you play with the robot nicely with the toys, it smiles.

If you overwhelm the robot by putting the thing in its face, it gets afraid and turns away. The idea is that not only is the nature of the stimulus important, but the quality, so the robot is trying to get you to interact in a way that is appropriate for the robot.

In the way that a parent would interact with a baby or a young child?

People typically put robots in hostile or different worlds [like factory floors or outer space]. It's very difficult for these robots to learn because the environments are always changing—you have unpredictable and imperfect information.

The whole idea in trying to work with Kismet is not to train robots in that environment all the time—especially if you imagine that these robots are going to interact with people. You can have a benevolent and teaching environment, much like people grow up in. So the idea is that the robot, instead of having to be a victim of having this complex world impinging upon it, can take a very active role in controlling that complexity.

It sounds very logical but not very easy.

This is a whole new way of looking at the big robot learning problem, by trying to draw on social and even cultural aspects. Anytime you have a person in the environment you have a potential for these kinds of interactions.

When you say "learning," are you talking about what an average person would understand by the concept of learning?

There are many, many different kinds of learning in any natural system. And Kismet certainly has exploited some simple forms of learning to do things like train ocular motor maps—like, if a stimulus is moving, how you move your eyes in order to engage it. That can be considered a very low form of neural learning that would happen in natural sys-

Peter's Notes

The first time I met Cynthia Breazeal I photographed her in Rod Brooks's Insect Robot Lab with six arms working on a six-legged robot for a story in *Smithsonian* magazine (see page 60). Two of her extra arms belonged to fellow student Colin Angle (in background on page 61) who today is still giving Rod Brooks a hand by running iRobot, his robot-building company. Cynthia still works hands-on on her robot head, Kismet—an amazing example of technical innovation, and perseverance. The first two times I photographed Kismet, the head was only capable of expressions after the application of a few dozen key strokes and the time it took the rack of computers to process the information. I had seen video of it working perfectly, but had not seen it in action. Then, on our third visit, the week before Christmas, Kismet was alive and well when we walked into the lab. The neck was craning around, the big baby-blue eyes were searching for something interesting. When I approached, they locked onto me and I had the robot's attention. But unless the baby is mine (unlike Faith, who can play with any infant for infinity) I get bored after twenty minutes with even the cutest little creature. After I ran out of stupid pet tricks, it was time to pass the kid over. Cynthia's Kismet was as cute as any baby I have ever played with, and I was thankful not to have to worry about diapers, but I was ready for a smarter robot after fifteen minutes.

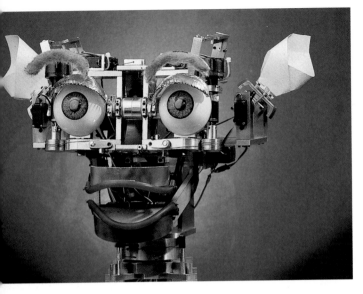
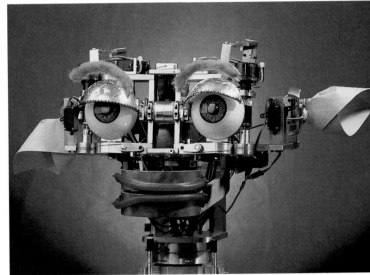

Kismet, the MIT face robot, can display a variety of human expressions (from left to right, surprise, disgust, tiredness, and the state of sleep).

Tech Notes

From Kismet's Web page
(www.ai.mit.edu/projects/kismet/)

Early Experiments in Regulating
Social Interaction

(Describing a series of preliminary tests performed with Kismet in 1998 involving human interaction, either through direct interaction—waving a hand at the robot—or using a toy to play with the robot.)

Two toys were used: a small plush black and white cow and an orange plastic slinky. The perceptual system classifies these interactions into two classes: "face stimuli" and "non-face stimuli." The face detection routine classifies both the human face and the face of the plush cow as face stimuli, while the waving hand and the slinky are classified as non-face stimuli. Additionally, the motion generated by the object gives a rating of the stimulus intensity. The robot's facial expressions reflect its ongoing motivational state (i.e., its "mood") and provide the human with visual cues as to how to modify the interaction to keep the robot's drives within homeostatic ranges.

In general, as long as all the robot's drives remain within their homeostatic ranges, the robot displays "interest." This cues the human that the interaction is of appropriate intensity. If the human engages the robot in face-to-face contact while its

(continued on facing page)

tems. That's way down at the sensory motor level, not at the cognitive or conscious level at all.

The kind of learning that I'd like to have Kismet acquire is more at the behavioral level. Can it learn to make a request for a certain toy? If it has learned to associate a vocalization such as the phrase "Can I have this?" with this toy and it does this vocalization, I give it the toy. To learn at that level, it has to acquire an understanding of people. The robot must understand people enough to be able to influence their behavior in a predictable and goal-directed manner. Ultimately, I want the robot to learn social behaviors and to understand people.

Is Kismet's learning cumulative?

We're still developing the learning software so the robot isn't learning anything right now, but the idea is that the learning would be cumulative. You can save the state of the robot if it ever powers down, and then ideally when it powers up you can start up where you left off. Ultimately, you are trying to teach your robot to learn from you.

If you look at it, there's a whole vast expanse of what you could be exploring. What I'm trying do now is really focus on the idea of using social interaction to resolve the learning problem in a new way. Because learning is so difficult in natural environments, because they are so complex, roboticists typically fine-tune and specialize their learning algorithms to teach the robot to learn this particular task in that particular environment. And so basically that's all that the robot can learn. It can't learn anything else because it's not been designed to learn anything else. In contrast, what I'm trying to do with Kismet is design a robot that can learn a lot of different things.

But instead of putting it all in at design time, which is what we do now, you would give the robot the ability to sense certain cues that can add the necessary constraints during its experience. Things like, if I want it to look at the block, I shake the block. If it has attention to motion, it will automatically look at this block. Then I might say something like "red" to draw its attention to the salient thing about the block.

Instead of the robot having to figure out on the whole visual field what is salient, I can help it to figure this out. That's a social cue, which naturally and intuitively helps the robot address the learning challenge.

The way a mother who would say to a child learning about color, "This is a red block." And in that way the mother uses social cues to help the child learn something. Are there other ways of helping it to focus itself?

Things like shared attention—where if I'm looking at this thing that's important and the robot knows where I am looking, it also looks at it. So, again, the robot is really focusing its attention on what I'm trying to help teach it. Or I could give it positive reinforcement. If it says the right vocalization, or something that sounds right, I say "that's good," or I smile. Give those kinds of social cues to guide its learning. Hopefully, by doing that you can have more flexible learning, so the robot can learn a variety of different things instead of just one thing the robot has been designed for at design time. So that is the big idea—to have a system be more open-ended.

Open-ended the way people are? You think this is the seed of the answer for robot learning—especially if these robots will someday be in our homes?

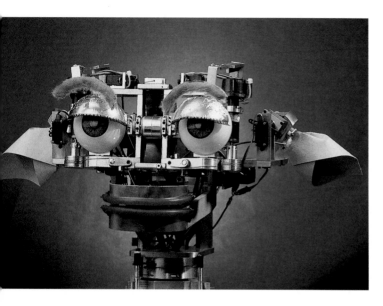

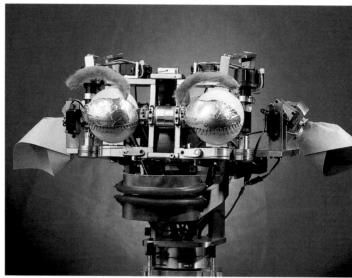

Certainly there are a lot of hard problems in traditional robotics as far as dealing with complex tasks in a complex world. I'm just trying to drive a different area, focusing on that social interaction. If you have a person in that environment and that person actually cares about having the robot do the right thing and learn, how can you exploit that kind of interaction to help learning?

You can imagine that if you have robots in your house, you are going to have to have somebody teach this robot something new, like taking out the garbage. They're going to work with this robot in order to show it how to do this and they're probably going to use the typical social cues they would use to teach another person. That's the natural, intuitive way people teach each other.

Which is what you're trying to harness—that natural, intuitive relationship. At this point, what is your relationship to Kismet?
I would be characterized as the caregiver, so to speak. The robot is like an infant—that's the whole scenario. The robot is what we would call an immature learner, at a lower level of sophistication than the instructor, who is like the parent or caregiver. The idea, of course, is that this is a tremendous opportunity for the robot, because this person knows a lot more than the robot does about where the robot is going wrong. The person has a sense of what is important for this robot to learn and so it can be a powerful influence in helping the robot learn the right thing.

You can imagine that if this robot were trying to learn from another robot, they might be having a hard time knowing what is right to learn, because neither really understands what's important. Whereas if you have this parent-child situation, you

have a very special kind of learning environment that I want to try to exploit with this robot. And once this robot learns something, then maybe it can be a teacher for a different robot. But really, what I'm focusing on right now is this human and immature robot learner scenario.

Presumably, not every robot has to be a social animal, so to speak. Do you get questions about that?
The field of autonomous-robot research started off with the idea that people were never going to interact with robots. They were doing tasks in specific environments and they had to have enough local intelligence to do that certain task. That was and still is a big motivating factor for autonomous robot research. More recently, there has been this push to think about, "What if you want people to work alongside these robots and interact with them?"

Now, interacting with them in a safe way, in a rich way, becomes important. And that seems to be one of the big driving motivators in humanoid robotics research. The argument being, sure, I can have a robot go around and empty my trash and deliver my mail but now I have to interact with this thing and communicate with it and get it to do new things for me. If it looks recognizable enough that I can use my set of social skills on it, and it's giving me the responses I would expect from another person, it is a much more intuitive way to interact with these robots.

Do you think that a robot could really interact in a way that felt psychologically natural to a person?
I think that the person who teams up with someone who really understands psychology or makes that an integrated part of the study is the one who will win the big social-robot race.

(continued from facing page)
drives are within their homeostatic regime, the robot displays "happiness." However, once any drive leaves its homeostatic range, the robot's "interest" and/or "happiness" wane(s) as it grows increasingly distressed. As this occurs, the robot's expression reflects its distressed state. This visual cue tells the human that all is not well with the robot. Then it is up to the human whether to switch the type of stimulus, or whether the interaction should be intensified, diminished or maintained at its current level.

Before the run begins, the robot is not shown any faces so that the *social drive* lies in the *lonely* regime and the robot displays an expression of "sadness."

During the next 3 minutes, depending on the stimulus, Kismet responds to small head motions, and "interest" and "happiness" appear on the robot's face. The experimenter begins to sway back and forth in front of the robot. This corresponds to a face stimulus of overwhelming intensity— first "disgust" appears on the robot's face, which grows in intensity and is eventually blended with "anger." Then the experimenter turns away, allowing the *drive* to recover back to the homeostatic regime and "interest" returns to the robot's face, followed by a face-to-face interaction of acceptable intensity, to which it responds with an expression of "happiness." Next, the experimenter turns away, causing the *lonely* regime and redisplay of "sadness," and finally re-engages the robot in face-to-face contact, which leaves the robot "interested" and "happy."

Getting Ahead

At the touch of a switch, the third-generation robot head in the Hara-Kobayashi Lab at the Science University of Tokyo flicks its dark eyes open and flashes a smile, showing a mouthful of pearly white teeth. Underneath its silicon rubber skin, "muscles" composed of narrow strips of a special shape-memory alloy flex in response to an electric current, creating an expression of happiness.

"Human-friendly robots seem necessary," says Fumio Hara, "if they are to coexist with us."

Like Cynthia Breazeal, who has built the robot Kismet at MIT (see page 66), Hara is trying to produce a robot with facial expressions that people can recognize. Both researchers believe that duplicating the physical attributes of the body may be essential to understanding human intelligence—and that people will be more likely to accept robots if they adopt human features. Beyond this, though, their approaches are quite different. While Breazeal focuses on using artificial-intelligence software in a physical body, Hara is much more concerned with the mechanics of facial expression—with creating an electromechanical structure that can precisely simulate joyful happiness.

Hara and his students have been building mechanical face robots since the early 1990s. The first attempt had a creepy oversized head—because the mechanism inside was bulky—with air-powered facial "muscles" that produced six rather clumsy facial expressions. At one point it was called Harumi, a common Japanese girl's name, although the robot's moniker has changed as often as the roster of students working with it. To make it seem more human, the robot wears a black blunt-cut wig and makeup. When we first visited, it was still operating. In fact, the lone female student in the group was daubing on its makeup.

Now, in time for our second visit, the third generation head is up and running. Human-sized and disconcertingly lifelike, it smiles with utter naturalness—but can't stop. Unfortunately, graduate student Masaoki Tabata tells us, the small strips of shape-memory alloy in its "muscles" return to their original state too slowly. As a result, the face sags back into its rest position after a smile like a time-lapse video of the force of gravity. The slow reaction, Tabata says, is an unavoidable tradeoff—the alloy strips are the means Hara's students use to shrink the size of the face-robot to ordinary human dimensions.

As we move into the lab area, Hara apologizes for the mess. There's no need, as far as we're concerned—the mess is photogenic. Ten or twelve students are milling about, working frantically. Here, as in other Tokyo universities, oral examinations are taking place.

Faith: You're working with social interactions between people and robots. But are you also programming the robot to be intelligent?

Fumio Hara: No, no, no. Most people try to write software that makes a robot do something, but my approach is somewhat different. We don't implement any intelligence-building software inside—just learning mechanisms and some kind of value system about what is good and what is bad for itself.

In the US and Europe the approach to robotics seems more software and artificial-intelligence oriented. They are working on the brain, so to speak, whereas you're working, literally, on the face—the physical structure.

I think the physical attributes of the body may be essential to understand our processing of intelligence. Usually people think about artificial intelligence as something inside a computer, but if we extend the concept of intelligence wider, or if we want to make some kind of universal process generalization, then we need to include the body when we think about intelligence. Computer intelligence is a subarea of intelligence itself. I'm not alone in this. Many psychologists are moving toward the idea that intelligence needs embodiment.

And a lot of roboticists are talking about the psychological aspects of robotics.

Yes, yes. For instance, if the face robot shows a somewhat strange facial expression to a human being, the human being will interpret the expression in a certain way. That may send the wrong message from the robot to the person, so the correctness of the expression we make is very important. The face robot's communication will become necessary at some future time, when robots have to take care of the older people or babies or some tasks like that.

The reaction would have to be recognizable?

Sure. There are two ways to make a face robot. One way is to use a cartoonlike face [like Kismet]; the other way is to make a realistic-type face, like our face robots. I don't know which way is better. At this moment we are trying to produce a realistic face robot.

But, you think we need physical embodiment in order for intelligence to manifest itself?

I think probably the shape of the robot or machine and the kind of material we use are important and have a strong influence on the intelligence. If the material or the shape is wrong then probably we'd need a lot of computation.

To compensate for using the wrong material?

Because if we have a very a rigid finger and want to hold it in a certain way, we need a very complicated computation. If we have a softer finger, we don't need so much computation to make it do what we want.

Our behavior is governed by the physical world because we are living in physical space. I am bound to exploit the physics in our interaction between ourselves and a robot or other artifact systems.

Robo Specs

Name: None—give her a good name, please.

Purpose: What happens when robots have humanlike faces is the biggest question ... as far as robot-human communication is concerned. Can robots have an "emotion" or not? Those two questions are the inspiration for our face robot.

Height: 360 mm

Weight: 2.1 kg

Vision: 1 CCD camera

Sensors: None

Batteries (type, duration): None

External power: 12 V (DC) power supplier, 300 W

KLOC: Not counted yet.

Cost: $85,000

Project status: Ongoing

Information from: Fumio Hara

Peter's Notes

Until I saw Fumio Hara's face lab in the basement of the Science University of Tokyo, the weirdest lab I had ever seen was a cryonics lab in Oakland, California. There I photographed the man who had been the world's youngest doctor (an M.D. at 18, he was then in his 30s) lying on an operating table reading *The Prospect of Immortality*. Behind him were gigantic stainless steel vats full of liquid nitrogen and severed human heads. Actually, that *was* weirder than Hara's lab. Hara's lab was merely crammed with skulls, model heads, computers, and face robots. At the workbench, grad students were applying makeup to an immobile, slightly sinister face robot—a scene that would have made Alfred Hitchcock smile.

A work in progress, this still-unnamed face robot (left) can open its eyes and smile. In the future, says its designer, Hidetoshi Akasawa, a mechanical engineering student working on a master's at the Science University of Tokyo, it will be able to recognize and react to human facial expressions. Believing that human communication is not accomplished by words alone, the laboratory has created three generations of increasingly sophisticated face robots in a program directed by engineering professors Fumio Hara (Akasawa's thesis adviser) and Hiroshi Kobayashi. This third-generation robot will greet smiles with smiles, frowns with frowns, mixing and matching six basic emotions in a real-time interaction that Hara calls "active human interface."

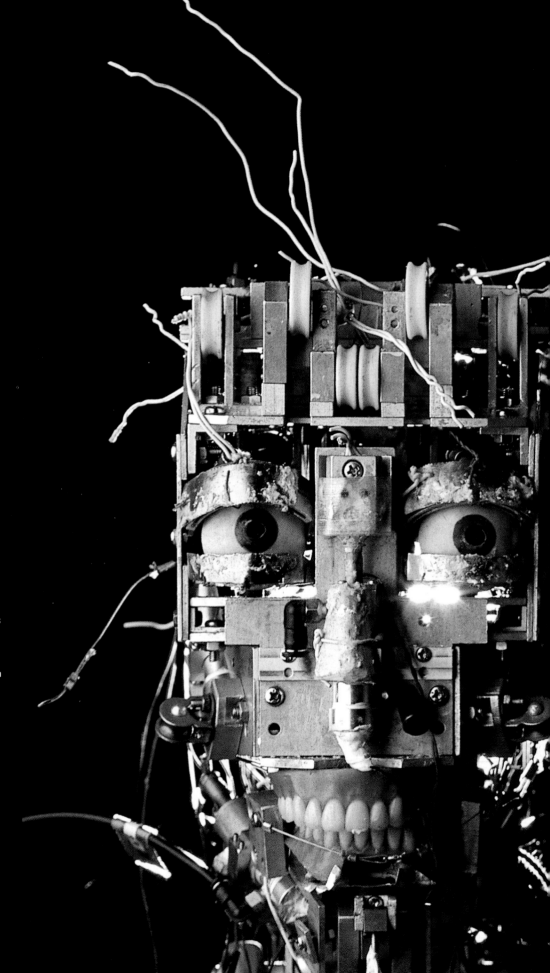

After he removes its skin, Fumio Hara gets the once-over from a face robot in the lab he co-directs with Hiroshi Kobayashi at the Science University of Tokyo. The first of several face robots made in his lab, it has a CCD camera in its left eye that sends images to neural-network software that recognizes faces and their expressions. Calling upon its repertoire of programmed reactions, it activates the motors and pulleys beneath its flexible skin to produce facial expressions of its own. The project is relatively unusual in its focus—many researchers believe that making robots walk and manipulate objects is so difficult that facial expressions are not yet worth working on. Hara disagrees, arguing that robots with animated faces will communicate with humans much more easily.

When Professor Hara showed us his first-generation female robot head, I immediately asked him to remove the sad-looking rubber skin and stringy black wig. As I suspected, the robot looked much more interesting with no skin and hair. When Hara's student, Hidetoshi Akasawa, showed us the translucent-skinned second-generation head, I asked him if I could light it up. "I guess so," was his puzzled answer. Once I taped a pencil-light strobe inside the head and showed him a Polaroid proof, he knew why. His scholarly vision of a cyborg head and my mutant vision of a stripped-down replicant melded in a flash. I wanted to take the head a step further—out into the *Blade Runner* neon part of Tokyo, the east side of Shinjuku. Professor Hara argued that the robot was fragile, was unfinished, and had never been out of the lab. Nonetheless, the next afternoon we picked up Hidetoshi and the head and drove to a parking garage overlooking the lights of Shinjuku. It had been raining nonstop for two days but we found a place on the sixth floor where we could prop up the head on a concrete wall and light it with a yellow-gelled pencil-flash. The rain damped the background colors, but we still got a cover image for this book.

Lit from behind to reveal the machinery beneath its skin, this second-generation face robot (right) from the Hara-Kobayashi laboratory at the Science University of Tokyo has shape-memory actuators that move like muscles creating facial expressions beneath the robot's silicon skin. Made of metal strips that change their shape when an electric current passes through them, the actuators return to their original form when the current stops. The technology has obvious promise for this application, but in this model the actuators were disappointingly slow to return to their initial state. As a result, the expressions seem frozen onto the face—one reason that the laboratory began work on a successive, third-generation face robot almost the instant this one was finished.

When we use physics, probably we don't need so much computation.

Can one instill the desire for life in a machine?

Probably, yes. But it may depend on the definition of life. There would be some analogy between the biological system and the so-called artifact system. If robots want to survive in a social environment—meaning, to be used by humans—then the robot probably will have to have some kind of function to adapt itself to human desires and wishes and hopes. That might be some kind of emotion, in a sense. Maybe in the future robots might have some kind of emotive function.

Do you think that a robot could at some point in the future experience fear, or hope, or longing?

Yes, I do. When we think about robot emotions, it would be very difficult for them to be implemented by a designer because the situations of robots and humans might be enormously different. You cannot account for all the aspects of human behavior that would be in the environment of the robot.

But if robot experiences would be so different from human experiences that their emotions would be inherently impossible to understand, what can you do as a researcher?

They might not be so much different from ours, because of social interactions with humans. My idea at present is that the face robot, for instance, has to learn how to behave through interactions with human beings—like a pet. If you are a good, kind person to the robot, the robot will become kind in return.

For some of the pictures, Peter removed the skin of the first-generation robot. When you do that, I would imagine that the reactions of humans looking at it are very different.

We haven't done this kind of experiment. But this face robot sometimes gives people strange feelings. We are now interested in why the face robot gives such feelings to human beings. That might be a good cue to improve our face robot to be more realistic. Right now, this robot can see your facial expression through her eyes and make a greeting.

What kind of response do you get when you show this robot?

Mostly, people don't seem scared—they just say "wow." When they saw the first one [the first-generation robot], they felt strange—some people were scared because its size was bigger. We feel fear when we see something huge like that. Even now when I see a very large machine on the street or in factories, I feel scared.

Do you think that your anxiety is unreasonable?

It might be reasonable if the thing were so huge that I thought it was going to destroy me. That's a good adaptation to the physical world if I want to survive as a biological organism.

Can your third-generation face do more than one expression now?

Yes, sure, in the near future. It only has one at this moment—but it's a very attractive smile [laughter].

Once the expressions are programmed, how quickly will they be able to change?

Almost as quickly as a human's. But because of the properties of the shape-memory alloy, when we want to return the expression to normal we will have to put some kind of small fan inside the head to cool it.

Will you give this one a voice box?

Yes, we want to implement a voice and include some emotional factors. It is difficult to produce an angry "yes" or a happy "yes." Subtle, quality communication through voice and facial expression is needed. If it's inconsistent, communication between robot and human will be destroyed.

Where do your ideas come from?

I don't know how ideas spring up. Take for instance a biological organism, a creature of nature. A robot is also a creature—an artificial creature. But if you look at both without any bias about their nature, you can ask, what are the basic principles to build such creatures? Both things have to be governed by the laws of physics. Otherwise they wouldn't function. So one answer is that even artificial creatures like robots have to be governed by certain principles to be able to have a particular body that can perform particular actions. That is my interest—the kind of principles we have to follow to make something.

How would you define yourself? As a roboticist?

I usually don't say I am a roboticist. If I say I'm a roboticist, I feel that it's not a perfect fit for myself. Just a scientist. Because in Japan roboticists don't usually pay much attention to the psychological aspects and I'm interested in the psychological aspects of this kind of artifact system.

There are futurists who say that man and machine will meld into one. What do you think about that?

I want to keep some kind of physical separation between robots and human beings.

There are already retinal and cochlear implants that, though they're artificial, are helping people.

Losing a hand and getting an artificial one is different from adding a mechanical improvement to a perfectly good appendage.

But there are people who will decide they want an implant to hear better, even though they have perfect hearing.

That might be okay.

Or a chip in the brain—a direct connection to the Internet.

At this moment, I think we don't know anything about what would happen if something went wrong. It's scary to think of a chip in a brain.

Tell me what you think of the Japanese National Humanoid Project, which is trying to create a robot prototype that will eventually coexist with people.

I don't understand what the goal of the Japanese Humanoid Project is—what they want to know by

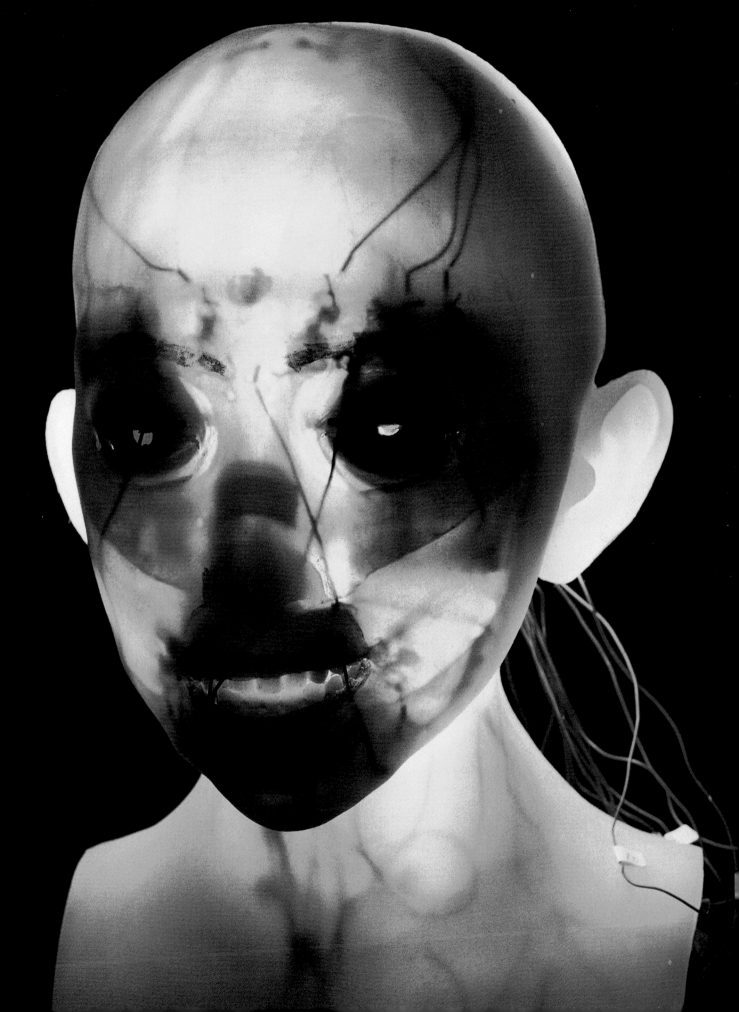

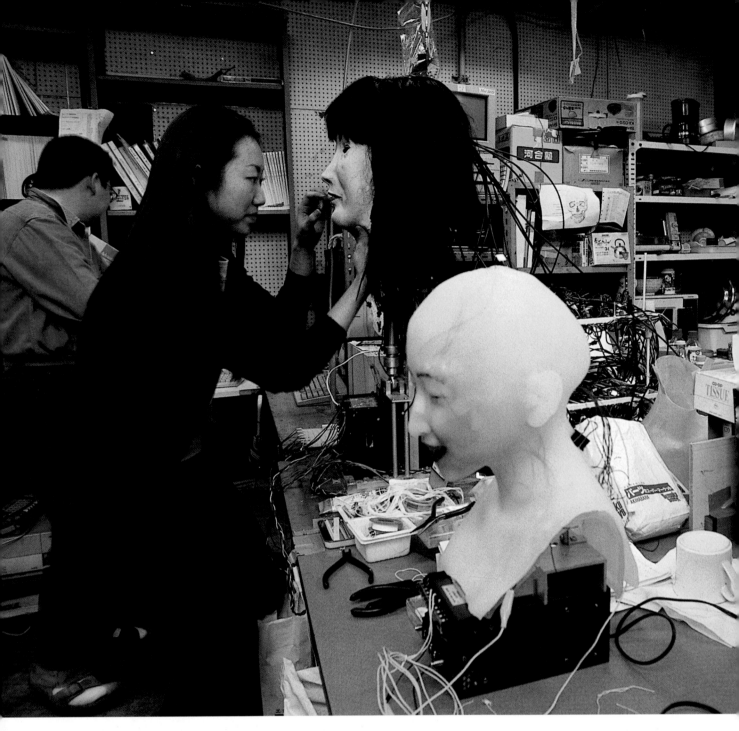

Trying to concentrate in a crowded, busy workspace, graduate student Harumi Ayai pats makeup onto the immobile features of a face robot in the Hara-Kobayashi Laboratory. This machine, the first face robot built in the lab, has a single camera in its left eye. (It is the same robot depicted without its skin on page 74.) Notwithstanding the relative simplicity of its design, the machine was able to smile when people approached it. Although rapidly superseded by later models—the lab went through three generations in a few years—the robot is still being studied.

making that kind of humanoid robot. Probably we could make such a thing by using very sophisticated mechanical-engineering technology. But what happens when we make it?

Do you know what they want the robot to be able to do? I understand that the government hopes this robot will be the cornerstone of Japan's future.

I don't know. I am not involved in that kind of project.

Why not?

The direction is quite different from my idea. When we build such a humanoid robot and the mechanical performance is very good then probably we will have some possibility to find a new field of science, but at this moment I can't see it. I stick to the face because

most of the information exchanged between humans is carried by facial expression and voice. Sometimes we use our hands, too, but not the body. The body carries a small amount of the message to us.

What about an elder-care humanoid robot, which is what many of the people involved in the humanoid project talk about? Could it work?

If we think about the type of care, we probably don't need a whole humanoid robot body to do it. And engineers will point out that if we make humanlike robots they will be too expensive—nobody deserves that kind of service. That is the engineers' viewpoint. So probably we need some kind of help, but we don't need a whole body to do it. If you want a robot to move, prob-

materials, and computation, with interaction in the physical world. You need them all.

The first-generation face robot was the thesis project of Fumio Ida who is just finishing his master's in the Hara lab. When I catch up with him, he is just about to leave for Switzerland to study at the University of Zurich.

Fumio Ida: Many people don't think about how robots can learn from human instructions. I think this is a very important topic, but it's very difficult because a human's response to a robot can be very personal or different from one another's.

Faith: Your face robot work deals with the interaction of humans and computers. What led you to come up with this interaction analysis?

I'm very interested in trying to determine the intelligence of a robot or a computer. On the other hand, I'm also interested in assaying the intelligence of human beings and how we get our intelligence. Originally, intelligence has a genetic source, but our educational environment is also very important. I'm focused on the development of the learning aspects of intelligence and that's why I focused on how the face robot learns intelligence from interacting with humans.

It learns? Do you mean learning in the way that humans learn?

At this moment she cannot learn as we do. But probably a few years from now she will, or some other kind of computer will learn as a human does.

Your lab director, Fumio Hara, thinks the shape and size of the body is critical to the way we learn.

Learning, or intelligence, is strongly dependant on the body, its shape—any kind of physical property. This is the very interesting research topic that I will study in Zurich. We call it "morpho-function"—morphology and function are very strongly related to each other.

Speaking of morphology, we spent some time with the Honda P3 robot. What do you think of it?

It is state-of-the-art technology. I think it's a good platform for a researcher—especially a human-interface researcher—because it is very difficult to build that kind of complex robot in a university or small laboratory or company. It's very expensive and difficult—our university couldn't build this kind of robot. If we had one, then we could implement our robotics into that robot as a platform.

Sony is giving their pet AIBOs to researchers, but there is a multimillion-dollar difference between the AIBO and the P3. What do you think about pet robots?

We are very interested in these. I'm interested because research on that kind of robot is very focused on human interaction with them. We don't have any idea what kind of usage or application a robot like this would have. I think this kind of research is interesting, but I don't know that it's very necessary for human society.

ably wheels would be better than humanlike legs. It is much more practical to build that kind of system. Especially when you consider that many elder-care environments are already set up for wheelchairs.

Correct me if I'm wrong, but I think a lot of mechanical engineers believe that if they are able to build a good body, the brain will come. Computer scientists think that if they are able to build a good brain, the body will come.

My solution doesn't favor either side. We need a body and a brain. Having only a body doesn't get you anything—it's just a machine. And something that is only a computer can't show any kind of behavior. The essence of intelligence is morphology [body shape],

Peter's Notes

Photographs are something I notice right away. In the Hara-Kobayashi lab at the Science University of Tokyo, behind the workbenches laden with tools, computers, female head robots, and skulls was an area where the students make tea and coffee and wash their dishes. Above the sink was a pinup of a girl in a red bathing suit. Somehow it seemed to fit the feel of the place. It reminded me of the seeing corporate-suited salarymen on the crowded subway trains nonchalantly flipping through girlie magazines. I didn't think more about this until we were back in California. Faith rented a cheesy black and white video called *The Stepford Wives*. In the movie a group of wealthy suburban men conspire to replace their wives with robot look-alikes who are not only subservient but fascinated with domestic duties. Back in Japan six months later we came across a story from *Shukan Taishu*, a girlie magazine salarymen read on the subway. The story was called "Unreal Inanimate Lover Has Guys Turning to Jelly" and it said a $7,000 life-sized sex doll called a "rarebot" was selling like hotcakes. The whole thing sounded bizarre enough that I had to grab a train to the suburban home of Masami Sakai, the designer of the silicon-skinned beauty. He came to the door and explained that he used to design prosthetic limbs. Then he got the rarebot idea and jumped right on it. Reona, the rarebot in his house, was wearing kimono pajamas and pantyhose. She slumped in a brown leather chair in the tiny office, wedged between a gold-framed mirror and a PC. Because Reona weighs as much as a person, Mr. Sakai has to cart her around in a wheelchair. To make a long story short, Reona was not a robot. All she had was a removable, washable insert that resembled a cross between a video cassette and a pet monkey. There were no motors or sensors, but I strongly suspect that someone, somewhere, is working on it.

Sleek and elegant, the head of
this unfinished robot was constructed by the Symbiotic
Intelligence Group of the Kitano
Symbiotic Systems Project. It is
funded by an ERATO grant from
the Japan Science and
Technology Corporation—a
branch of the Science and
Technology Agency of the
Japanese government. SIG, as
this robot is named, has a white
outside shell designed by a project artist—group leader Hiroaki
Kitano is a firm believer in the
importance of aesthetics. The
stress on elegance paid off: SIG
is the only robot ever invited to
be displayed at the prestigious
Venice Biennale art show.
Though it is scheduled to be
completed sometime in 2000, it
will be a descendant of SIG that
eventually becomes a freely walking humanoid about the size of a
child. The small size is no accident: Kitano thinks that people
are more likely to accept robots
that look too little to hurt them.

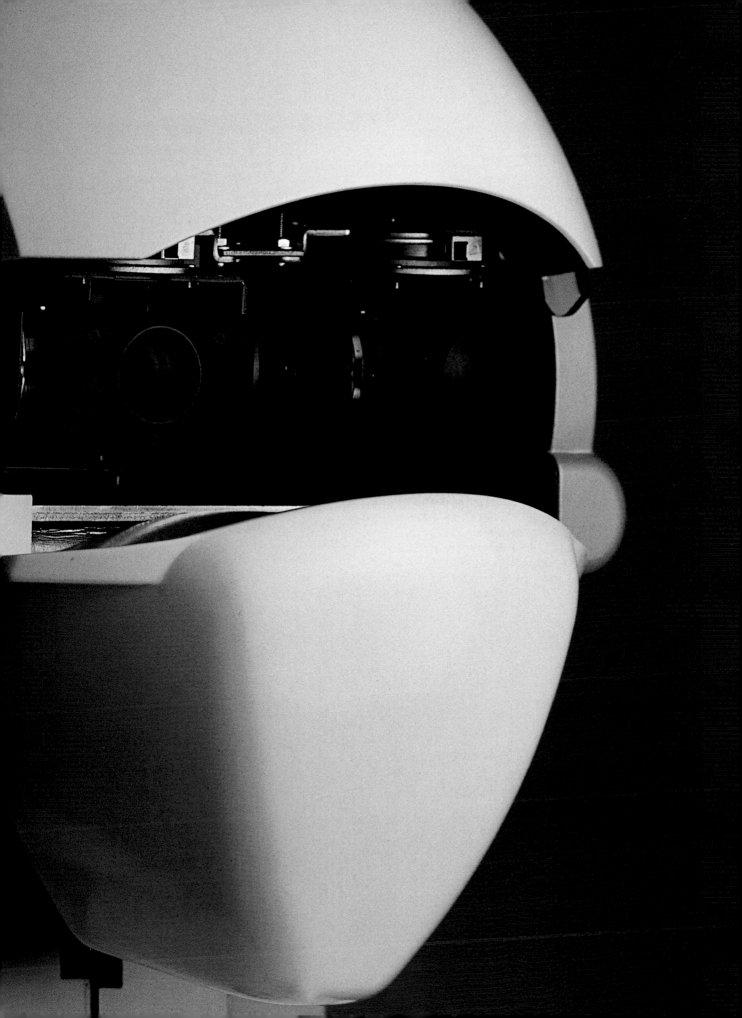

SIGnature Robot

Robo Specs

Name: SIG

Origin of name: Symbiotic Intelligence Group (name of project research unit)

Purpose: Study intelligence on multimodal sensory integration on high-DOF [degree of freedom] robot...also will serve as a prototype for upper-torso part of full-body humanoid.

Vision: Digital camera

Sensors: Four microphones for stereo hearing

Frame composition: Aluminum

Computers: ERATO-1 Beowulf-class 32-PC cluster (6.7 Gflop/s)

Project status: Ongoing

Information from: Hiroaki Kitano

Peter's Notes

Style means a lot in Japan—in calligraphy, clothes, cell phones, cars, computers, cameras, even... robots. Hiroaki Kitano's new offices are in the heart of tony Harajuku, wedged between designer boutiques, coffee houses, and trendy restaurants. Condomania, Japan's biggest condom emporium, is only a few blocks away. On weekends, Elvis and Elvira clones flock to a corner of Yoyogi Park and young women with huge platform shoes giraffe along the asphalt sidewalks. Kitano's office is tasteful—glass walls, hardwood floors, an iMac and an AIBO (Sony's robot dog) in the reception area, white slippers for guests. On our first visit, with the droll enthusiasm of an Inspector Gadget, Kitano showed us his RoboCup robot soccer team: miniature electric whizzers with pinball flippers, and a goalie that orients its position by aligning with magnetic fields. His team did not do well at the RoboCup world championship in Stockholm in August 1999. But Kitano was undaunted. A month later he invited us to photograph his new humanoid robot, which was designed by an inhouse artist. As soon as I saw it, I said to Kitano, "It's like Robo Cop. You're going from RoboCup to Robo Cop. It's beautiful." Kitano smiled, and agreed. His goal, he explained, was to build the "archetype" humanoid robot. He said he was creating a new field called robot design. I am betting he will succeed.

Hiroaki Kitano is not the only researcher in artificial intelligence who thinks the future will include humanoid robots that can think on their feet and operate in the real world. But he is one of the firmest believers that corporations had better be prepared to see them come on the market—whenever that might be.

When we encounter Kitano, he is helping Tatsuya Matsui, his artist-in-residence, with a confusion of papers and computer cables. Carefully balanced atop the pile is SIG, the new humanoid robot of the Kitano Symbiotic Systems Project in Tokyo's Harajuku District. Being little more than a shell at this early stage of development, SIG travels light. But not for long—Kitano has big plans for this robot. Matsui is only beginning to grapple with the task of fitting the hardware into SIG's shiny plastic head.

Director of this Japanese government-sponsored research project; senior researcher at Sony Computer Science Lab; chairman of the RoboCup Federation—an association for international robot soccer—Hiroaki Kitano is a busy guy. What this entrepreneurial scientist enjoys most seems to be whatever he happens to be working on at the moment. And at the moment, he's focused on SIG.

Kitano has hired Matsui because SIG is the first step of his robot design initiative. He's looking beyond the maturing of robot technology, which he takes for granted, to what he calls the eventual entrance of humanoid robots into the human world. SIG, he says, is also his research team's platform for the lofty goal of "eventual uncovering of the underlying principles of intelligence."

Balanced atop the pile of papers, SIG at this moment is just a shiny robot façade of white molded plastic with a smoky visor over eye-level cameras. It has no arms and legs—they will come later, promises Kitano. The group intends to build a full-bodied, albeit short, humanoid robot. I ask him why he worries so much about how the robot will look, even at this very early stage.

Hiroaki Kitano: Basically, the reason we are working on the design is that industrial robot design will be critically important when there is a huge market for humanoid robots or other robots with many degrees of freedom.

Faith: As important as design is to the automobile industry?
Auto design is a major area in industrial design. We want to create the prototype for all robot design. Anyone who builds a humanoid will have to come back to this humanoid. It's going to be the archetype. *It's very Star Wars. When will the physical body be finished?*
We're going to first build a skeletal model with all the motor components in there. Then we'll test it

out and probably we'll throw it away and remodel it [laughs]. Then we should have a second version, with Matsui's external design completely integrated. A humanoid robot is incredibly hard to engineer and there are significant safety issues if the robot is going to be walking around in our world.

I think that a biped humanoid shape has an advantage in terms of accessibility and mobility in our environment. There is the argument that the humanoid robot is unstable at this moment and a wheel-based robot might be more safe. But I think that stability will dramatically increase over the next ten years. It's just a matter of time because we know that we can build a humanoid. The fundamental part is done. *Is that the time period you think—ten years?*
This takes time but I think that the first humanoid robot will be sold in the year 2003, within the current scope of technology.
Who's in a position to do that?
Some Japanese company, maybe Honda—but I am just guessing. 2003 is a very special year for the Japanese. You know about Astro Boy? *Tetsuwan Atomu—the Japanese robot-boy cartoon. It ran on American television in the 1960s. I know it's very popular here.*
In the story, 2003 is the year when Astro Boy was created. Everyone working on humanoids in Japan is working toward 2003.
You'd better get working! We had a conversation the other day with Masato Hirose of the Honda P3 project [see page 43]. I got the sense that they're trying to figure out what to do with their robot.
Who is going to buy it? It's not very obvious right now. It may be that the technology is not mature enough to be commercialized within a few years. I don't think anyone knows the correct answer to that. *Is it a good robot? Some people say it didn't break new ground technologically—not that it had to, I guess.*
It's old, conventional technology that uses zero-moment-point computing as a control technique. But they actually built it, and other people didn't. And they spent the money! That deserves some recognition. *Would you have built the same kind of robot with that time and money?*
What I'm trying to do is to create a humanoid robot, but without using explicit computing, as they did—telling the robot exactly what to do. Humans are not programmed explicitly. We'd like to have a more humanlike neural control system, where the robot actually learns to walk and its walking behavior gets better and better over time. Honda's robot is not going to get better by training.
It's only as good as its programming. Whereas yours, in theory, can get better and better?
Yes, but if I had been in their position ten years ago, I might have done what they did. I think it was a practical choice to use conventional technology to engineer something that could walk, and see how far they could

go with it. Given the fact that they started ten years ago, it was a very reasonable thing for them to do.

Can they build on what they've already built?

I think so. What they should do is make it much smaller, more compact, and then try to sell it in a limited market and see how people react to it and use it.

As Sony is doing with the robot pet AIBO [see page 224]?

We were not quite sure how people would play with AIBO before we started selling it. Now we have a very good understanding about how people react to this kind of robot; who is going to buy it, and how they're going to play with it.

Do you have to spend money these days in order to build a good robot?

I think you have to spend quite a bit of money.

Can a university compete at a high enough level?

I think university robotics research, except for very advanced theoretical things or very novel ideas, can only survive for the next ten or fifteen years.

But isn't that the best place to learn the fundamentals?

Who's doing automotive research right now in universities? Who's building semiconductors? For edu-

cation, it will continue to be a valid topic. But, once an autonomous robot industry is formed, options for university-based research will be limited.

People have already tried and failed a few times to set up a commercial robotics industry.

There are a number of robotics companies starting out. Some of them will grow and some of them won't. And also, the big companies like Honda and Sony are jumping into robotics with their tremendous capital. Generally, though, the argument I would make to a university researcher is to stay away from that competition, because there is no way the university can win. Go for something more advanced, that industry cannot afford to do.

But is that in engineering or in artificial intelligence?

I think it's more in AI. Or engineering that industry cannot do because it's too futuristic or risky. Like if a researcher says, "In the future, artificial muscles will be created from hundreds of little microdevices," a company would say, "Ohhh, not yet."

Do you see yourself as an AI guy?

I think so. I'm not really strictly a robot guy. If a robot has no intelligence then I'm not very interested.

Surrounded by his plans and sketches, designer Tatsuya Matsui (above, seated) contemplates the next phase in the evolution of SIG, the robot under development by Hiroaki Kitano (standing). Kitano, a senior researcher at Sony Computer Science Laboratories, Inc. and director of this government-funded project, wants to endow SIG with sufficient eyesight, hearing, and processing power to follow instructions given by several people in a crowd. The goal is ambitious, but Kitano is well-placed to achieve it. In 1997, he created the now-famous RoboCup, in which robot teams from around the world meet every year to play soccer in an indoor arena. Though intended to be fun, the competition has a serious purpose. Kitano, mindful of the universal appeal of soccer, thinks that people who have watched robots "play" will feel more comfortable with these metal entities.

Bio*logical*

Bi·o·mi·met·ics, *n.*, the use of living creatures as inspiration for machines [*hist.*, ~1995, *Amer.*]

Just below the surface of a reservoir outside Boston, robot Ariel walks sideways like the crab it is patterned on.

Aping Behavior

Peter's Notes

Toshio Fukuda's lab was as no-nonsense as the city of Nagoya that housed it. The rooms were packed with robots of every description, from bipedal machines and monkeylike brachiators to bunches of tiny devices that interacted like insect colonies. The human element consisted of bleary-eyed students all dressed in their Sunday best—our visit coincided with the last day of oral exams. Professor Fukuda was the most gracious harried host I have ever met. Although late for his own banquet, he insisted on walking us across the campus to be impromptu guests of honor. We wanted to stay much longer but we bowed our way out at 10 PM for our five-hour drive to Tokyo. He understood—we had to shoot more robots in the morning.

Hanging from a network of cables, Brachiator III (right) quickly swings from "branch" to "branch" like the long-armed ape it was modeled on. (Brachiator refers to "brachiation," moving by swinging from one hold to another.) The robot, which was built in the laboratory of Toshio Fukuda at Nagoya University, has no sensors on its body. Instead, it tracks its own movements with video cameras located about four meters away. Brightly colored balls attached to the machine help the cameras discern its position. Brachiator's computer, which is adjacent to the camera, takes in the video images of the machine's progress and uses this data to send instructions to the machine's arms and legs.

Harried, personable Toshio Fukuda leads us on a breathless tour of his robotics laboratory at the University of Nagoya, where his projects run the gamut from small autonomous devices intended for space exploration to child-sized walking humanoids (see page 46). What snags our interest most is Brachiator III, a robot that swings, monkey-style, from handhold to handhold. Like most laboratory projects, it is not quite ready for the real world, but nonetheless it swings agilely through an artificial course, moving almost silently, and is astonishingly adept.

Brachiator is a puzzle. Roboticists have long tried to create robots that look and act like human beings: and almost always, the results have been meager—stiff, slow-moving machines that nobody would ever mistake for living things. By contrast, Brachiator, though clearly nonhuman in its appearance, darts through the metal "trees" almost as gracefully as a living creature. (When it's wearing its custom-built monkey suit, the machine is eerily lifelike.)

Brachiator's rapid progress exemplifies the radically different approach that some roboticists, Fukuda among them, have recently begun to employ. Instead of being guided by the physics they learned in school, these researchers are drawing inspiration from new discoveries in biology. Known as "biomimetics" or "biomechanics," the approach is a fusion of insights from nature and mechanical engineering.

In biomimetics, biologists feed information about organisms to robotics engineer; in turn, the engineers give back data about the physical parameters that characterize natural systems. Both disciplines collaborate to build robots—biologists as a means of experimentally verifying their understanding of natural principles, engineers to use those natural principles to create machines unlike anything ever built before.

Underlying biomimetics is the engineer's growing appreciation of evolution's astonishing ability to innovate—and the biologist's similarly increased appreciation of the possibilities of engineering to extract information from living organisms. Using a battery of novel, high-tech methods, researchers are beginning to learn exactly how animals move. Some are looking at the dynamics of motion; some at its neurological basis; and still others at the abstract physical principles embodied in anatomy. Over and over again, they are discovering that even such a seemingly simple phenomenon as how fish swim involves complex—and previously unknown—physical principles.

Some scientists are actually attempting to replicate organisms exactly—researchers at Case Western are developing a robot cockroach that duplicates every bone, muscle, and nerve (see page 102). But that is not the usual approach. Evolution produces organisms that grow, mate, and reproduce; robots, built for limited purposes, do not need to do any of these things. As a result, biomimetics researchers usually build robots to imitate only some aspects of living things, ignoring the rest. An example is the iRobot company's Mecho-Gecko, whose innovatively designed feet are inspired by the unusual features that let a real gecko stick to the wall. The idea, Fukuda and other scientists say, is for engineering to do what nature does—but to accomplish it by means that nature cannot.

Faith: What was the reason for building Brachiator?
Toshio Fukuda: It's a good test bed for enhancing the learning and adaptation capabilities of a truly dynamic robotic system. It's immediately recognizable, even without its fur, which was essential for people to understand it immediately. It's the responsibility of advanced-robotics researchers to show the importance of their research to the public sector, including funding agencies.
Does it have any practical applications?
Yes and no. There are no direct real applications, but it is being sponsored by industry to represent the state of the art for learning and adaptation technologies. It's a demonstration that can be applied indirectly to industrial processes.
What do people say when they see it for the first time?
"It's simply fantastic!" When the robot wears the monkey suit, it really looks alive.

Roboticists took a surprisingly long time to realize the power of biological models. In the late 1970s, Shigeo Hirose of the Tokyo Institute of Technology, built a robotic snake (see page 89). Its fluid movements startled his colleagues, but they didn't fully embrace the idea until the mid-1980s, when Marc Raibert of the Leg Laboratory at the Massachusetts Institute of Technology created a series of robots that ran, hopped, and even turned somersaults. The contrast could not have been greater between Raibert's agile creations and the awkward, barely functional machines that were then the stock in trade of robotics.

Today, biomimetics is at the center of a worldwide explosion of robotic innovation. Bob Full, at the University of California at Berkeley (see page 90), is studying the physical structures of living organisms to create robots that can negotiate the complexities of the real world. Alan DiPietro of the iRobot company (see page 93) is focusing instead on the properties of the nervous system and the brain, which can modify their own behavior to adapt to changes in the environment. Mark Tilden, of the Los Alamos National Laboratory (see page 117), is using biology to create machines with behaviors that his colleagues find almost impossible to explain. To Fukuda, the implications of this creative explosion are clear. At the current rate of advance, he says, in fifty years a network of robots "will be all over the world."

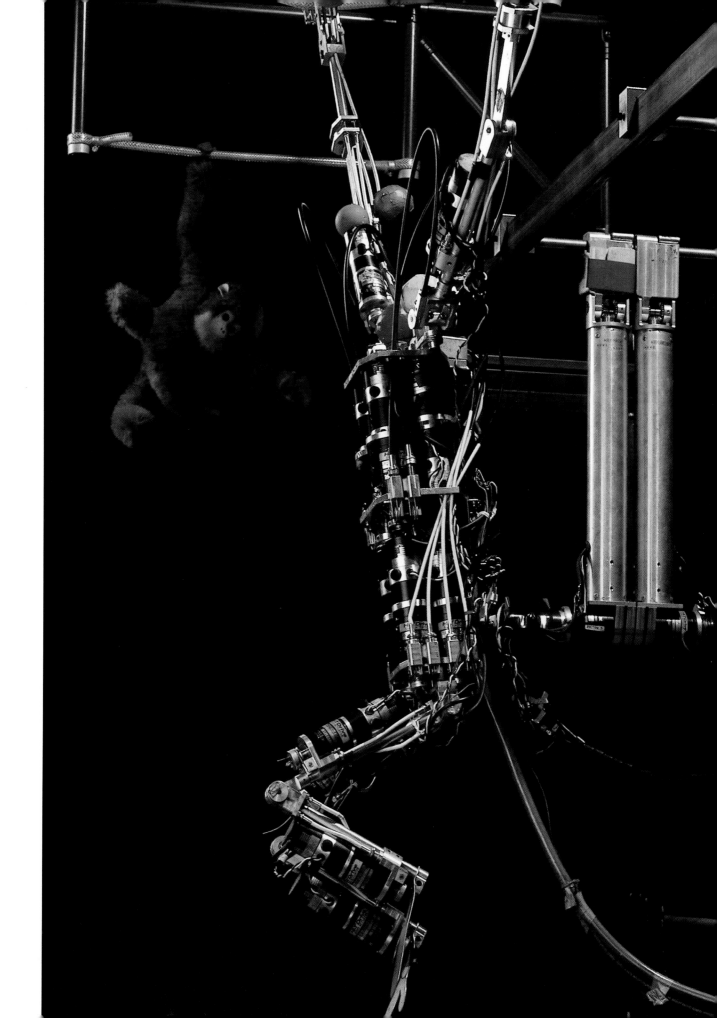

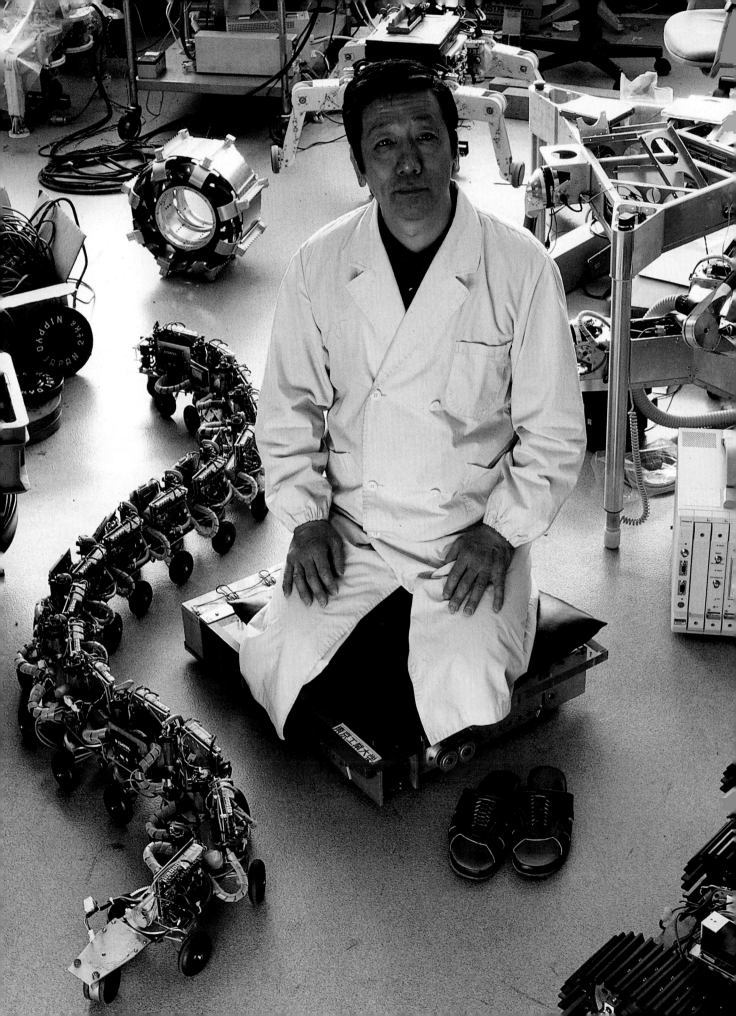

Snake Charmer

The name of Shigeo Hirose at the Tokyo Institute of Technology comes up early and often when engineers say which robot builder has most inspired them. In the mid-1970s, years before most roboticists had even heard of the words "biomimetics" or "biomechanics," much less contemplated their meaning, Hirose was studying the locomotion of snakes and using his findings to create wheeled robots that undulated across the ground. His fascination with the simplicity and efficiency of the snake extended to other creatures as well; he went on to create four-legged walking robots derived from his research on spiders, and rolling machines that somersault, tail in mouth, along the ground the way stomatopods roll across the ocean floor.

At the time, the idea of drawing on evolution—rather than engineering—for inspiration was startling. Hirose's argument that progress could come best from new and untried methods seemed to fly in the face of experience, but today Hirose's work is regarded as groundbreaking and the number of biologically inspired researchers is growing.

The change in attitude occurred in part because roboticists have come to understand how difficult it will be to create humanoid robots. By contrast, Hirose's machines work—they do what they're supposed to do. Hirose's lab is a formal place, reflecting his formal demeanor. It's packed with his creations—which represent a career of three decades in engineering. While his assistants carry out some robots for Peter to photograph, I ask him if he built things when he was a child.

Shigeo Hirose: From the time when I was very young, I liked to make things. I was interested in radios, so I made a radio set. Once I made a pneumatically driven alarm for a Japanese bath. The [Japanese tub] is a little bigger than a Western-style tub and it takes more time to fill. So I built an alarm to tell when the tub was full.

Faith: No chance of it overflowing then. Your mother must have been happy about that.

Yes. When I was in high school I had to concentrate on my studies. But sometimes when I was very busy, just before examinations, I would think of a new idea and work on that instead. That was my frustration. But now, when I go to my laboratory, I'm very happy for I can keep thinking. It is a kind of paradise for me.

You've said that you're an engineer, not a roboticist, even though you're a robotics pioneer. What's the distinction?

I'm an engineer. My objective is to make useful things. And if those useful things aren't robots, I'll still want to make them.

You've also indicated that you don't think much of humanoid robots. Why?

An engineer is one that makes very useful things and sometimes what is made is very simple. If a roboticist sees a man washing clothes in a river, he would try to make a humanoid that could approach the river and scrub the clothes. If an engineer sees this, he would make a washing machine, a very simple rotating machine.

A lot of researchers, especially in Japan, are trying to build a humanoid just to see if it can be done. What do you think of doing it for that reason?

At a conference I attended in the United States, many people from Japan supported building humanoids. But I thought that it was very important to say that there is a different opinion, even in Japan. So I said—rather strongly—that to build a humanoid for the sake of building a humanoid seems to me kind of silly.

That was brave—telling a whole bunch of your colleagues that their life's work was worthless [laughter].

I believe that if we want to create something useful, we should have maximum freedom to select its shape. To create something useful within the restrictions of the humanoid is very difficult and ineffective, so I don't like it.

Do your students ever tell you that they want to build humanoid robots?

Yes, of course—most of the students do. When I ask for their ideas about robots, they're thinking about humanoids. But I convince them to make something different [laughter].

Do you think it's even a bad idea to try to use robots to mimic human thought?

Yes. Another bad point about the humanoid is that we try to imitate the human mind with them. The human mind is a product of evolution, but a robot that we make ourselves should be very different. It won't be a creature that is always trying to stay alive—and most immorality comes from that. Robots can keep all of their information in [computer] memory, and if they are completely destroyed we can make a new mechanism and restore their memories.

So humanoids won't have to worry about struggling for survival.

People fear robots because they think they will rule human beings. But the reason we ourselves want to rule other people is that it's much easier to survive when we are in a position of power. But robots can act unselfishly. Robots can be saints—intelligent and unselfish. For humans, being intelligent is easy, but being unselfish is rather impossible.

Did you expect, when you first started as an engineer, that it would come to the point where you would have to be thinking about these things?

For now, the level of robotic intelligence is getting higher and higher. It is approaching our way of thinking a little. We have to be very serious in considering this point. Until now, it was a rather abstract concept—it's not so important, but in the future, I think it will be very important.

Robo Specs

Name: ACM R-1

Origin of name: Active Cord Mechanism, revised model

Creative inspiration: Observation of snake.

Purpose: Off-road vehicle, new manipulator.

Length: 2 m (20 joints)

Maximum slithering speed: 40 cm/s

Cost: The series took 29 years, more than 60 students. ACM R-1 took half a year with 1 student.

Project status: Ongoing

Information from: Shigeo Hirose

Name: Vuton

Origin of name: *Zabuton* (Japanese cushion for use on *tatami* mat)

Purpose: Multipurpose, omni-directional vehicle

Height: 13.5 cm

Weight: 29.5 kg (without batteries)

Maximum carrying capacity: 1,090 kg

Project status: Ongoing

Information from: Shigeo Hirose

Sitting on a mobile motorized cushion he calls a "vuton," Shigeo Hirose of the Tokyo Institute of Technology (left) surrounds himself with some of the robots he has built in the last two decades. Beside him is the snake-bot ACM R-1, one of his earliest projects. It is made of modules, any number of which can be hooked together to produce a mechanical snake that slowly, jerkily undulates down its path. Hirose, who is primarily funded by industry, hopes to develop commercially useful robots; the snake, he thinks, could be useful for inspecting underground pipes. He is designing three new, improved models. The vuton, too, may be marketable. The robot cushion uses a system derived from the snake—a circular, bulldozerlike track created, in essence, by joining the snake's head to its tail in a continuously moving roll. Controlled by a joystick, the vuton may help older people move about—something that could be important in a country with a rapidly aging population.

The Full Story

Name: Mecho-Gecko (see page 92)

Purpose: Provide mobility to small-sensor package, extend range of operator's senses by climbing to high vantage point.

Creative inspiration: Gecko lizard's climbing and attachment abilities.

Height: 6.3 cm

Length: 11.4 cm

Weight: 118 g

Vision: None

Sensors: None

Frame composition: Delrin

Batteries: Lithium primary

External power: None

KLOC: 0

Cost: $600

Funding: Defense Advanced Research Projects Agency

Project status: Ongoing

Information from: Alan DiPietro

Peter's Notes

Every time I visit the Poly-PEDAL lab at U.C. Berkeley, Bob Full has something he is really excited about and wants me to see. He is like a very smart kid enjoying the heck out of a field trip to the zoo. But Bob's zoo is more like a circus, a menagerie of little show-offy creatures: a ghost crab that skitters sideways like an eight-legged ballerina on speed; a cockroach running an obstacle course that suddenly sprints on its two hind legs like a runner out of electrified blocks; high-speed close-up videos of a running gecko's feet attaching and peeling off of a transparent vertical treadmill. He loves to share what he learns about animal locomotion with roboticists, and gets a huge kick out of the cool machines they build using his information. During my last visit he was *really* excited. Tiny artificial muscles from the Stanford Research Institute were being tested next to a real cockroach's leg muscles on a workbench. Professor Robert Full was in cockroach heaven.

One of the most interesting and influential people in robotics today is a biologist—Robert J. Full of the University of California at Berkeley. He is at the center of several groups of roboticists—even though he neither builds nor designs robots himself. Instead, he studies the principles of animal locomotion, a subject about which surprisingly little is known. Cheerful, enthusiastic, impatient, blessed with a remarkably intuitive understanding of his subject, the biologist spends his days filming, testing, measuring, and dissecting a remarkable variety of creatures, but he is perhaps best known for his examination of cockroaches and geckos. Building on his cockroach studies, researchers at Stanford and the University of Michigan are creating multilegged, insectoid robots; Full's gecko work has helped inspire iRobot, a robotics firm in Somerville, Mass., to construct a geckolike machine that can crawl over walls and ceilings (see page 92). Yet although Full helped pioneer the transfer of biological principles to robotics, he is among the first to criticize the misuse of the idea. First, biology was ignored, he says, then it was wrongly embraced. In fact, Full believes that although biologically inspired robots are beginning to realize their promise only now, they will make enormous progress in the next few years. I asked him what he thought was wrong with the way roboticists have used biology.

Bob Full: If you take a look at many of the robots you've taken pictures of, and the several hundred multilegged robots that seem to be inspired by creatures, what you find is relatively minimal performance in terms of speed, stability, and maneuverability. Why is that? What's the problem? I believe that looking at nature for answers to design has been applied by many engineers in a completely inappropriate way.

Faith: Is it because the engineers don't understand biology?

—And many of the biologists don't know how to express it to the engineers clearly. It's a communication problem. I wouldn't say it's anyone's fault. It's led to a very specific problem. When you have a task at hand—you want to make a mobile robot or have some kind of behavior—and you look toward nature, do you blindly copy it? Do you emulate it, or not? The classic answer by engineers and even some scientists who don't understand evolution very well is, "Millions of years of evolution has evolved something, so it has to be pretty spectacular." Of course, that is not true. Organisms have

historical constraints. They have ancestors and they can't throw away all their genes.

So if you want to make a robot that walks like a human being, you don't have to copy the poorly designed lower back and knee we've been saddled with by evolution.

The real secret is to understand the fundamental principles and transfer them to the engineering. This is very challenging—you have to know something about the physiology and biomechanics of the organism, something about development, ecology, evolution, and behavior. At the same time, do what you do best, engineering-wise. Don't give that up.

Nature isn't always the model?

I don't see why you can't build things better than nature will ever have thought of. Planes already show it—I mean, you can't ride on a bird. In some ways, engineering has just blown away nature and in other ways it's terribly inadequate. There is no robot at present that can run up that wall in a matter of seconds.

Whereas your gecko can do that easily. As a biologist, what's your role in robotics?

We provide biological inspiration to design—that's the goal. Provide that kind of information to the best engineers in the world, to get them to build things that don't exist now.

Are you interested in robotics, per se?

Yes, but I'm far more interested in understanding how creatures work. Nothing can beat having a creature and trying to figure out how it moves. When you discover the secret, and for that instant nobody in the world but you knows it or has ever known it. And you know it.

Give me an example of when that happened to you.

The first time we found out that legged creatures all seem to work like spring-mass systems. Their legs compress, store, and return energy, and that produces a pattern of forces. That was startling—we knew then that some day that idea would lead to a highly mobile robot.

How can springlike legs make robots more mobile?

A spring, in a sense, is a controller. If you stretch it more, it pulls back harder. It doesn't require a sensor to send a signal somewhere, which then responds to that signal, to pull itself back. It frees you from the constraints of sending out a signal, sensing it appropriately, getting it back, and acting within the time that you need. That's what the organisms do. They blast over irregular terrain at incredibly high speeds with no way to make the instantaneous calculations of where their foot is and how to respond to its location.

Most of the robots we've seen have powerful computers to tell their limbs where to move.

For novel behaviors, you need the onboard processing to tell every joint at every instant what every other joint is doing in order to move forward.

Walking slowly over irregular surfaces is a very complex control problem. If you hit a bump, you have to figure out that you hit the bump. Then you have to send a signal back to correct for hitting the bump, and as a result you end up being slow and not very maneuverable. Your sensors can make mistakes and you have to have circuitry that's coordinated correctly. Humans can do it—and other animals can do it—for slow behavior. For novel behaviors, you have to have that kind of springlike control, otherwise you can't execute them.

What we are seeing in fast, running animals is that you just don't need to do all that. We discovered that if you're a sprawl-postured animal—your legs are splayed out to the side—and you have many legs, then you can put the control in the mechanical system rather than in the neural system. Instead of having little motors in each joint, where you are sensing everything and carefully adjusting them, you put a spring there that provides instantaneous feedback.

It's beautifully simple. There's no sensors, length or force transducers, there's no neural circuitry. Essentially, the legs are doing the computations on their own.

Give me an example. You study cockroaches—how do they run?
It's the same principle for all legged animals. They bounce when they run. Many-legged sprawl-postured animals were thought to only walk. We discovered they bounce too, but do it like Groucho Marx, with bent legs and no aerial phase. Running does not require all legs to be off the ground.

Basically what's happening is that they use alternating tripods. Three of their legs [two on one side, one on the other] all operate together and work like one of your legs. Then the next three come down and work like your second leg. And they bounce along like that.
Bent down low, bouncing forward while swaying from side to side.
The American cockroach does leave the ground because at fast speeds it runs on only two legs at fifty steps in one second—it's really fast! If humans could run that

Gloved to ward off the possibility of a nasty bite, Berkeley biologist Robert J. Full (above) prepares to pluck a gecko from his office door. A source of inspiration to roboticists around the world, Full's Poly-PEDAL laboratory is one of the premier research centers in the field of animal locomotion—that is, how living creatures move around. (Polypedal means "many-footed"; PEDAL is an acronym for the Performance, Energetics, and Dynamics of Animal Locomotion.) Full has studied the movements of a variety of animals: ghost crabs, ants, cockroaches, beetles, fiddler crabs, centipedes, geckos, and stomatopods (underwater crustaceans that somersault along the bottom). His work has played a major role in expanding the world of robotic biomimicry, giving inspiration to the makers of Ariel (see page 101), RHex (see page 96), and the "sprawl" robots (see page 98).

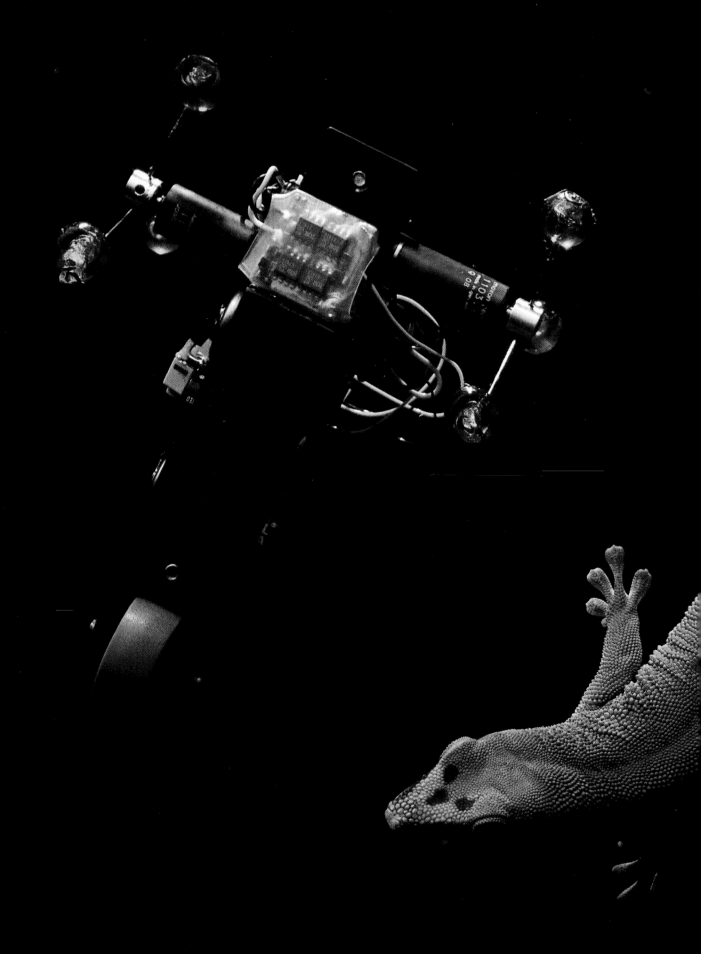

The tips of the gecko's toes are covered with corrugations of fantastic complexity—it's as if they had the world's most complex fingerprints, with the world's most prominent grooves and ridges. Actually, the corrugations are lines of tiny hairs. Flattened in the right way against a surface, the hairs lie so tightly on the surface that the gecko's toes literally form a kind of chemical bond with it. (In technical terms, the gecko takes advantage of van der Waals force.) Planting their feet down and peeling them up,

Hunched over a treadmill designed for arthropods, biologist Robert Full tests an Arizona centipede in his laboratory at UC Berkeley. Even though the centipede has forty legs, it runs much like an ordinary six-legged insect. Just as insects move on two alternating sets of three legs (two on one side, one on the other), the centipede gathers its legs into three alternating groups, with the tips of the feet in each group bunched together. Full analyzes centipede motion by observing the insect's movement across a glass plate covered with "photoelastic" gelatin (above, lower). On either side of the gel are thin polarizing filters that together block all light coming through the glass. When the centipede's feet contact the gel, they temporarily deform it, altering the way light goes through it—and allowing some to pass through the filters. In the test above, one group of legs works on one side of the animal's midsection while two other groups work near its head and tail.

fast, they would go two hundred miles per hour.

You also study geckos—how they can skitter around on the walls and ceiling. How do they do that?

Pick a toe and watch it very carefully. The toes peel back like tape. They do it in milliseconds. They do it without any forces you can measure. We can physically view it now with better technology, but we couldn't see it before because it happened very quickly.

The toes are sticky like tape?

There are 500,000 tiny hairs on a gecko's foot. Each of those hairs has fine branches, with a billion ends that have intimate contact with the surface.

If the hairs were small enough and the hair was orientated properly to the surface, you could create a molecular bond between the hair and the surface—a van der Waals bond.

The forces are enormous, in fact, so big that if they all attached perfectly—which they don't—then one gecko foot could hold a basket with a hundred apples in it. But, the gecko can get away with using ten percent of the hairs and still do well. They can

deal with any angle, because they have so many hairs that some will be laid down properly. You can push them any way and they grab on. But once the toe is down, it's stuck, and then they have to peel it up to take it off. You can see it in the video—all these little tiny ends that have attached.

Now you're working on artificial muscles—you're actually ready to test them?

Not the artificial muscles that you hear about all the time—that people say are artificial muscles—but muscles with characteristics that fit within the functional space of real muscles.

What are the characteristics of real muscles? Don't they just push and pull the body around?

Really, what the muscles seem to be doing is controlling the stiffness of things—modulating stiffness. That makes perfect sense if things are just springs, bouncing along, and you don't want to get in their way. That's what we find the muscles are doing. They're kind of acting like springs and shock absorbers. This is a whole different way to think about the muscular skeleton.

It seems like such a simple thing that wasn't known.

There are twenty-one muscles in the cockroach "knee." Why are all those muscles there? What are they doing exactly? The first time I looked in a book and saw all those muscles, I said, "What do they do?" We still don't know that. You can tell what they do in the grossest sense, then you look at the animal and you see that the muscle is not even active—

Are all the science books wrong?

I would say that the science books are correct if you think about how isolated muscles work. In and of itself that muscle works that way when you take it out of the organism. The books are wrong when they tell you how you use that muscle when you run.

What are the implications for understanding this kind of thing, apart from their implications for robotics?

The biggest problem in biology is complexity. It's terrible. You have an organism and it has way too many neurons for what you think it does. It has way too many muscles, hundreds of muscles—why?

Way too many joints to do almost anything you can think of, and in many cases way too many legs or appendages. Why does it have all these things? You have this very complex system and you can't figure out what in the world the advantages and disadvantages of that design are. So our work is a demonstration of how you can take that complexity and simplify it to say, "You have all those joints but it works like one leg, and you have all these legs but it works like one virtual spring."

You can simplify it down to a little point-mass and a spring. The idea is to take that complexity, simplify it, find general principles of operation, use that as a target to control things, to figure out what is going on, and say, "That's what that leg does."

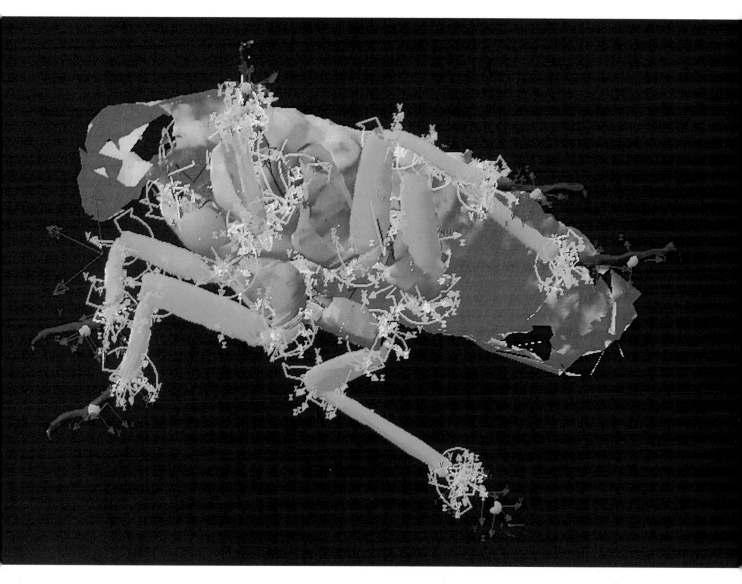

Then you go to the next level. If that's true and I have all these legs and all these joints, how come I have all these muscles? What are they doing? That's what we're trying to learn.

At a lot of robotics labs, researchers don't seem to share research with each other easily, and if patents are involved, forget it! But you collaborate all over the place.

I don't have any proprietary issues because I'm not building the things. I'm suggesting design ideas. When we've discovered something that they want to patent, we try to stay out of that. We give up stuff because what we care about is communicating these cool design ideas.

You're having too much fun to worry about the intellectual-property issues?

Things are changing so much that if you don't do this you won't progress! There is no way in the world that one person is going to make big leaps now. If you don't collaborate with people outside your discipline you will never make it. It's a really big jump.

I didn't think you could get any more excited than you were about your work the last time we spoke, but now you are even more caught up in it.

This is incredibly exciting! This is a culmination of ten years of me saying, "This will work if you just try it!"

You were talking about this ten years ago?

Absolutely. That was the first time we discovered this. The new advance in the last two years is that we found self-stabilization occurring. Fifteen years ago, we knew the bouncing mechanics for sprawl-postured animals with many legs. In the last few years we quantified it and in the last year we mathematically defined it with the help of the Koditschek group's mathematicians [at the University of Michigan] and Phil Holmes at Princeton.

Now I'm certain we're going to have swarms of these cheap things—toys first. In five years, you will have small dynamic robots that are capable of performing all sorts of activities. The main purpose we are thinking of is search-and-rescue. They can very quickly zoom in to find people in burning buildings without firefighters dying. And in ten years really spectacular things will be available.

Shown in color on a computer screen (above), this representation of a cockroach from below encapsulates years of measurements in the laboratory of biologist Robert Full. Full puts roaches on treadmills, measuring the electrical impulses in their muscles with tiny electrodes and photographing their movements with high-speed cameras. To ascertain the energy an insect's muscle generates, he stimulates the muscle and measures force as it is oscillated through the changes occurring in the animal when it runs. The research is summarized in the image, which is complete with arrows indicating the possible movements of each joint. The simulation of the cockroach on the computer screen was created by Mariano Garcia.

Road Rhex

Robo Specs

Name: RHex 0

Origin of name: Robotic Hexapod —Zero because it's a proof-of-concept prototype.

Purpose: To demonstrate that our ideas could be used to build a new class of legged robots, even with only one actuator per leg—a fraction of most other hexapod robots' actuators—and yet achieve unprecedented mobility over varied terrain.

Creative inspiration: Watching a movie of a cockroach [by R.J. Full] racing over rough terrain. In contrast to legged robots built to date, it didn't seem to do any careful planning of footholds, didn't avoid bumping into things, and generally just kept thrashing forward.

Height: 15 cm (26 when standing)

Length: 53 cm

Weight: 7 kg

Vision: None

Sensors: Battery voltage, battery current, motor angles

Frame composition: Aluminum

Batteries: Two 12 V 2.2 Ah lead-acid

External power: None

KLOC: 5.6

Cost: $10,000 (components)

Project status: Ongoing, DARPA funded

Information from: Martin Buehler

A surprising amount of the lab's work at Robert Full's Poly-PEDAL laboratory at UC Berkeley focuses on cockroaches, because they are exceptionally mobile—for their size, the fastest species on the planet. The fastest roach is a big species known, melodramatically, as the death-head roach, seen here in its "run" (right) at the Poly-PEDAL lab. As the run demonstrates, cockroaches do not have to have secure footing to move quickly. Instead, they use two alternating sets of legs (two on one side, one on the other) as springs, almost bouncing themselves forward. Remarkably, the insect brain doesn't have to see its feet or even be aware of them— an insight that Martin Buehler of McGill University and Daniel E. Koditschek of the University of Michigan seized upon when they created RHex (far right, controlled by graduate student Uluç Saranli). Tested in a laboratory dominated by an antique poster for Isaac Asimov's book, *I, Robot*, RHex could become a "companion robot," Buehler says, following its owner around like a friendly mechanical shadow.

With its decal eyes, fake plastic antennae, and mouse-pad feet, RHex isn't much to look at. But the conceptual underpinnings of this six-legged robot are a potential blockbuster for robotics.

In 1998, Martin Buehler, a McGill University roboticist, was captivated by biologist Robert J. Full's video of cockroaches navigating a bumpy surface. The bugs moved at terrific speed, their legs churning almost too fast to see. But in ultra-slow-motion videos, Full revealed that the creatures were constantly stumbling and bumping into obstructions. Nonetheless, they continued to run swiftly, suffering little loss in performance.

With Full's encouragement, Buehler decided to borrow the principles of cockroach design to create a walking robot. According to his longtime mentor, Dan Koditschek of the University of Michigan, Buehler roughed out the design by sticking "paper clips into an eraser—a conceptual eraser—and said, 'We can build this.'" RHex the roach-bot is now a full-blown collaboration among Buehler, Koditschek, Full, and Michigan graduate student Uluç Saranli.

When Buehler shows off RHex he cautions that it is just an early prototype. But even in its crude, unfinished state, the squat-bodied, autonomous machine is wicked fast. Set in motion, RHex moves so quickly that its legs are hard to see, much like the cockroach's legs in Full's video. "What's next?" I ask. "Where do you hope to go from here?"

Martin Buehler: Let's say that RHex 1 [the finished version of RHex 0] does really well, that it hops and runs and does more eye-popping demos. RHex 2 will have articulated legs, with just the knee, to be able to get better mobility. Right now the mobility is somewhat limited—it can't go under cars, because it has to rotate the legs around. So it should have more knee-like, more anthropomorphic design. After that, RHex 3 would have even more actuation per leg—have a hip and a knee—so it would look more like a cockroach. There are a couple of options that are not ready yet for prime time, but in two years, they might be. At that point, we could actually be building a high-performance robot with articulated legs.

Faith: Like Case Western's roach robot [see page 102]?

Sort of, but with novel actuator technology, so that we wouldn't pay all of the weight and size penalties of traditional actuators. Its unlikely that we could do that with the kind of motors that we use on RHex 0 here.

Poor Rhex is only a zero.

It's just an initial, backroom prototype.

Robots that are closed-loop do something, sense the reaction, modify their actions according to the feedback, sense the result, and so on. But RHex is open-loop—it can't sense the reactions to its motions.

This open-loop nature—as crude as it seems—in a way captures some of the essence of what's going on in this cockroach as it races at high speeds over fractured terrain where it has to move its legs faster than its neural circuits can actually control. A lot of real cockroach behavior is probably open-loop or is closed with a very fast feedback path around the legs. Which is really what started this whole thing at the meeting—seeing this cockroach going so fast, with the suspicion that perhaps inside there is very little closed-loop going on. It thrashes its leg forward in a tripod gait, which is really what this does.

When Saranli slows the robot down, I can see what Buehler means by "tripod gait." Each leg moves in a circle—somewhat like the way swimmers sweep their arms around to do the crawl. The front and rear leg on one side and the center leg on the other side move together, so the robot (like the roach) has two coordinated groups of legs. The legs in a group all cycle to the ground at the same time, thrashing the bug forward. Then the next group of legs comes down, and then the first, and RHex relentlessly churns and bounces ahead.

"My only concern is that one of its legs might break, and we don't have a back-up leg on hand," Saranli says. He's really only concerned because of our photograph. Part of the reason for building RHex—and other biomimetics machines—is that their components are cheap. RHex's limbs are little more than bent plastic rods. If the robot breaks a leg, the researchers can replace it at any hardware store.

Happy Feet

Robo Specs

Name: Mini-sprawl

Origin of name: Smaller successor to "Sprawl," so named for its sprawled posture; the "mini" prefix is taken from the movie *The Spy Who Shagged Me*.

Purpose: Test basic locomotion principles with an easily reconfigured robot in a variety of real environments. We also wanted to identify critical areas in small robotics that can be improved by shape deposition manufacturing (SDM).

Creative inspiration: The cockroach, because of its ability to traverse rough terrain in a fast, robust manner. We are not interested in exactly copying the cockroach in its physical form. Instead, we want to extract the basic principle of fast and robust locomotion that exists in many animals.

Height: 10 cm

Length: 15 cm

Weight: 260 g

Vision: None

Sensors: None, which is actually important.

Frame composition: Hardboard

Batteries: None

External power: Pressurized air—pneumatic

KLOC: 2

Cost: $330

Project status: Ongoing

Information from: Jorge G. Cham

Peter's Notes

In January 1990, *Forbes* magazine asked me to shoot an assignment at AutoDesk, which is best known for its computer-assisted design (CAD) software. "Something new, crazy, very California," promised the photo editor. It turned out to be virtual reality. After the shoot, I strapped on the goggles and data glove and flew through walls and into a swimming pool. I immediately sensed what Silicon Valleyites call TNBT—The Next Big Thing. I felt the same way in January 2000 when I visited Mark Cutkosky's lab at Stanford. He made CAD designs of robot legs and automatically created molds from the designs. Then he embedded small motors and sensors in the molds and poured flexible polymers around them, yielding remarkably flexible robot parts—imitations of bone and muscle. The process is called shape deposition manufacturing: SDM. SDM = TNBT.

Mark Cutkosky's robotics lab at Stanford University is transforming the design of robot components. Unlike conventional robots, which have a myriad of wires, gears, cables, and other parts that can break, loosen, or go out of adjustment, Cutkosky's robots have sensors and motors embedded in lengths of plastic polymer. Made by a process called shape deposition manufacturing (SDM), these integrated components—flexible structures with embedded electronic "nerves," "bones," and "muscles"—have no mechanical moving parts. (They are what might be called solid-state robotics.) The simplicity and durability of SDM components have allowed researchers to create a new variety of biologically inspired robots. Mimicking sprawl-postured animals, these small robots are able to move their six legs in a dynamic hopping motion patterned on the motion of insects.

Starting first with individual modules, such as leg joints, Cutkosky's group is progressing to entire limbs, and will eventually create entire SDM robots with flexible, ultra-sturdy bodies and limbs. The Stanford team is collaborating with Robert J. Full's group at the University of California at Berkeley (see page 90) and with physiologists at Stanford, Harvard University, and Johns Hopkins University. Although Cutkosky's progress toward building full SDM robots is slow, the newest small hexapods in his lab—a series of "sprawl" robots—have evolved at breakneck speed. In less than a year, Sprawl was succeeded by Mini-sprawl, which was succeeded by Sprawlita. When Peter and I came by, Cutkosky showed us Mini-sprawl, which was built by his graduate students Jorge Cham and Sean Bailey. At first glance, the paperback book-sized robot had the slapdash look of a high school science project. But as the pneumatic robot danced across the workbench, I began to appreciate what Cham and Bailey had done. As I spoke with them, Mini-sprawl exhibited a serious case of happy feet—all six of them.

Faith: Do you remember the first time you realized that you wanted to work with robots?

Jorge Cham: I think I knew as a kid that I wanted to build robots, maybe partly due to cartoons and comic books.

Any particular one?

Sean Bailey: Transformers definitely had an influence on me. I was always a tinkerer—my mother stopped buying me toys because two weeks after they'd buy one I would take it apart.

Why did you build this robot? You did it in two days?

Cham: To build the actual robot, yes. We had already spent time on the software and the electronics for the bigger robot, which took about two weeks.

Bailey: We were looking through a lot of the robotics literature, and we were seeing all of these really complicated robots that were very limited in what they could

do because they were so complicated. Bob Full has been telling us that cockroaches really aren't that complicated. They rely a lot on simple things like passive mechanical properties, sprawled posture, geometry, et cetera, and yet they can do amazing things like crawl over fractal surfaces three times their body height. We're trying to figure out why cockroaches have those capabilities and why the traditional robotic approach isn't working. So, late last spring we started to really examine cockroaches themselves. We had a couple of cockroaches in our lab, death-head cockroaches from South America. They are rather large—about three centimeters long. We started really looking at the kinematics and how legs were arranged on the body, and we came up with the simplest mechanism that could duplicate what was going on in the cockroach leg.

Were you experimenting as you went along?

Cham: That's another main reason why these things are put together so quickly. It lets us change things very quickly—see what works, what doesn't work—instead of some robots out there in which the whole thing was designed, built, and then it doesn't work so well.

Are you planning to add computers to control it? Or, beyond that, try to use artificial intelligence?

Cham: I think that it is in the plan of our project, once we figure out how the mechanical system works, and how it works with different surfaces and terrain. Then we can introduce things like learning and adaptability into the robots, so that they can handle different kinds of terrain or obstacles.

Bailey: That is what Reza Shadmehr at Johns Hopkins keeps drilling into us—he's one of the collaborators—that high-level control is a lot easier when you have good low-level mechanisms working.

What's your goal for the next Sprawl?

Cham: Fast and robust.

The robustness is from your embedding of actuators and sensors in one compliant piece—more like animal bodies?

Cham: What we're trying to do now is a shape deposition manufacturing version of Mini-sprawl. You can imagine where the servos [see page 235] will be embedded in this body—one here, one there, and one there. Just like Mini-sprawl, except it will be encased in this plastic material, very robust. In addition to the body we are also making legs, from the same idea. The servo arm will be embedded in the plastic, a flexible piece, and then the pistons will be embedded in the plastic, so it will be much more robust.

Do you enjoy the collaboration between biology and mechanical engineering?

Bailey: Research in the future is going to be collaboration. It's not going to be sitting down with one problem and working at it until you get it done. It's going to be collaborating with researchers in other fields and figuring out the results of the collaboration and what you can achieve.

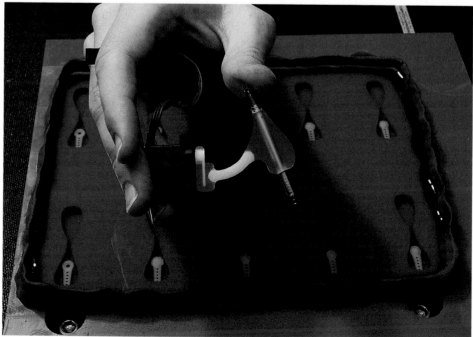

Rather than building an exact metal and plastic copy of an insect's bones and muscles, Stanford engineer Mark Cutkosky and his students Sean Bailey and Jorge Cham (above, Cutkosky at left) stripped a cockroach to its essence. The Mini-sprawl has padded feet, with springy couplings and pneumatic pistons that yank the legs up and down. Like a real roach, the robot skitters forward as each set of legs touches the surface. The next step: creating a robot that can turn and vary its speed.

Holding what will become a robot leg, Stanford graduate student Jonathan Clark (left) demonstrates the structure's resilience. Using shape deposition molds like the one below Clark's hand, Cutkosky and his students are now embedding electronic parts into molded plastic to create structures with the flexibility of living tissue.

Crab Dip

When the robot stops walking underwater, Ed Williams, a technician at the Massachusetts robotics firm iRobot, wades into the reservoir and plucks it off the bottom. The catch of the day is a machine called Ariel, modeled after a crab but designed to hunt mines, both underwater and on land. Completely invertible, the battery-powered, autonomous beast has no "right side up," which means it remains functional if the surf flips it over. Housed in a waterproof body cavity, the electronics feature self-adaptable software—if one of its six legs fails, the software realizes it and adjusts to a five-legged gait. The military hopes to have fleets of these bottom crawlers in communication with each other to thoroughly sweep an area for mines. Today, though, there's only one Ariel, and it isn't able to communicate with anyone. To demonstrate the machine to us, iRobot programs it to walk for two minutes in a nearby reservoir and then shut down. This way Ed knows he won't have to use his diving mask and fins, as he did the time Ariel got away from a TV crew and started crawling around the center of the reservoir on its own.

Faith: It walks pretty well underwater. Do you think it could cross this pond?

Ed Williams: Probably not. Right now, we can run for an hour on batteries. I doubt we can walk across the lake in that time. But battery technology simply has to get better.

The next step is fuel cells.

Yeah, more battery life. That's the key to the whole thing. We're not nearly as efficient as animals are, in terms of taking in energy and transmitting it back into power.

What are the two things on either end? Are they sensors?

Yeah, they are two identical devices. They have a compass and also a tilt-and-roll sensor. The reason they are so far out there is that they are away from the motors—the motors have incredibly powerful magnets in them. That far out, they're not affected by them.

Software is very important with an autonomous robot. How smart is Ariel?

You can do a lot of artificial intelligence with a really big controller. If you wanted to throw a Cray [supercomputer] at Ariel, you could do incredible things.

Or Deep Blue, the chess-playing supercomputer.

But no good chess playing robot has been able to scramble down a rough beach. For dealing with real-world situations, coming up from the hardware and sensors is very much how animals do it.... By having a very small controller, and really bringing that mindset into every single robot that we make ... we can do things that are fairly successful and fairly tight, in cheap packages.

You seem to enjoy the process—

We just printed up a bunch of T-shirts with our company mission on them: "Make money. Have fun. Build cool stuff. Change the world." Real cool stuff. The change-the-world thing is important. We do go home saying, "Gee, we can make these things and clean mines off of the beach." If we make a ton of money, great. If we don't make a ton of money, we'll still retire with a pretty good smile.

Robo Specs

Name: Ariel

Origin of name: Main character in the movie *The Little Mermaid*.

Purpose: To find mines.

Creative inspiration: Biological inspiration from crabs. Engineering inspiration from limitations of previous robots.

Height: 9 cm, stands 15 cm

Length: 55 cm (main body); 115 cm (including outrigger arms for compass and inclinometers)

Weight: 11 kg

Vision: None

Sensors: Motor position, foot contact, pressure (underwater), flow (underwater), compass, inclinometer, metal detectors

Frame composition: Anodized aluminum

Batteries: 22 NiCd cells

External power: Optional (while recharging batteries)

KLOC: 6

Cost: $50,000 (materials only)

Project status: Ongoing

Information from: John Aspinall

Sidling along the edge of a reservoir outside Boston, Ariel the crab-robot (far left) moves with a slow, steady, sideways gait. A machine with a serious purpose, it is designed to scuttle from the shore through the surf to search for mines on the ocean floor. Ariel was funded by the Defense Advanced Research Projects Agency and built by iRobot, a company founded by MIT robot guru Rodney Brooks (see page 58). Inspired by research on crabs at Robert Full's lab at Berkeley, Ariel takes advantage of the animal's stability—and improves on it. Unlike real crabs, which must struggle to right themselves if a wave flips them on their backs, the robot simply reorients itself and keeps walking with its body upside down. But despite its abilities, the technician in charge of the machine, Ed Williams (left), supervises Ariel's excursions with great anxiety—the machine still gets stuck when it encounters big rocks. "Robots can't do much now," he says, philosophically, "but airplanes couldn't do much in 1910."

Roach Approach

In an old brick laboratory at Case Western Reserve University in Cleveland, research associate James Watson plucks a death-head cockroach from a bin full of the insects and immobilizes it with a blast of carbon dioxide. Working with tweezers, he glues tiny electrodes onto the leg muscles of the animal, then marks the joints with white paint to make them more readily observable when he puts the cockroach on a tiny treadmill. As the glue dries, he tells us what he's doing—we've never seen cockroach fitness training before.

Watson works with Case Western biologist Roy Ritzmann, who is collaborating with Roger Quinn, a Case Western engineer, on a robot cockroach. He will match his data on leg-muscle signals to measurements on motion from the video of the running cockroach. The engineers in Quinn's lab will then try to replicate the muscle movements in the robot roach. In an unusual effort, Ritzmann and Quinn are building a mechanical insect that looks and acts precisely like its biological counterpart. Their current robot, Robot III, built by graduate student Richard Bachmann, is a rather ungainly pneumatic cockroach robot, seventeen times larger than a real roach but otherwise almost identical in form. Robot IV is in the works.

"Robot III is a little behind on the walking," says Quinn, as Ritzmann and he take us to see it. I soon see what he means. Although the mechanical insect occasionally jumps suddenly, frightening passersby, it can barely take a step forward. Having watched a live roach charging along on Watson's treadmill, I ask Quinn and Ritzmann how well their robot can walk.

Roger Quinn: As far as walking goes, it has walked, but really badly and that's because we didn't have strain gauges on the legs at the time. [Strain gauges measure and compensate for the stresses on the legs.] I mean, walk badly in that it tended to lurch and look like it was a jumping cockroach, maybe [laughs]. We've got the strain gauges on, and now we're developing the controller software so that we can make it walk again. It should be able to walk nicely.

Roy Ritzmann: Robot III is capable of capturing the kinematics of walking for the cockroach. It is not capable of capturing the dynamics of walking because the weight properties are not scaled properly because of the actuators—the pistons are very heavy. *Faith: In lay terms, the robot can reproduce the roach's movements—its kinematics—but it can't reproduce the forces involved—the dynamics— because the actuators are much heavier than muscle?* Quinn: Robot IV will be an attempt at getting the dynamics right. We decided to do the kinematics first and then move on to the dynamics. And with Robot IV we will use actuators that will actually be inside the leg, which also gives them the possibility of putting strain gauges on the leg that are more animal-like. *The strain gauges will let you compensate for forces, which is something that muscles do. Is this an area where your work overlaps a bit with Bob Full's at Berkeley [see page 90]? He, too, is trying to figure out the dynamics of muscles.* Ritzmann: Bob is much more interested in the muscle out to the skeleton and has the expertise to measure those forces. Our expertise is more from the nervous system to the muscle. We chose the cockroach that we work on because he had done a lot of data on it. We do talk to each other regularly, and it turns out that our stuff jibes pretty well, but he tends to concentrate more on fast moving, and we tend to concentrate more on slow moving. *The cockroach in your study actually moved pretty quickly even with all those electrodes hanging from it.* Ritzmann: When the animal starts going really fast, those wires we are recording these potentials from start to clump together. From Full's perspective, he's likely to push the idea that the sensory reflexes aren't doing a lot, and that it is the dynamics of the muscles and the skeleton that are principally responsible for the roach's movements. That is probably more true when the animal is going fast rather than slow, although we think it's not completely true. Some of [the difference in perspective] has to do with the fact that I'm a neurobiologist and he's a dynamicist. *So you focus on the reflexes, which involve nerves, and he focuses on the dynamics, which is his field of study. What are you attempting with your collaboration?* Quinn: What we are trying to do is solve the problem of locomotion. We want to build a robot that is like the animal in as many ways as possible, so that we can actually make a robot that is as capable as the

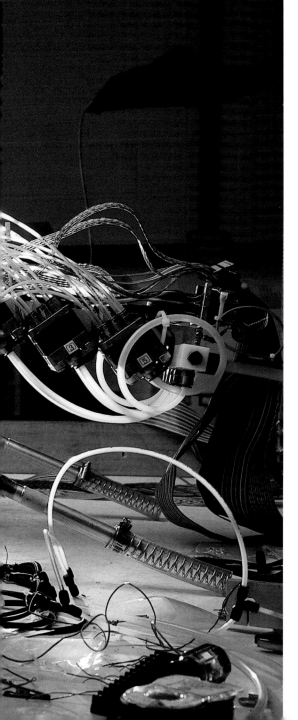

Robo Specs

Name: Robot III

Origin of name: Third in a series of hexapod robots. Our goal is to develop robots with the remarkable locomotion capabilities of cockroaches. Rather than starting with a robot similar to a cockroach, like RIII, we started with a much simpler hexapod, RI. With each robot in the series we solve more insect locomotion problems and gain the knowledge to proceed to the next robot in the series and each robot is designed to be progressively more similar to the cockroach.

Purpose: To build a robot that is strong and powerful enough to carry out missions and that captures the leg designs that the cockroach uses so successfully for running and climbing.

Creative inspiration: Leg geometry and joint designs are scaled from *Blaberus discoidalis*, the death-head cockroach (robot is 17 times larger than its animal model).

Length: 76 cm

Weight: 13.6 kg

Vision: None

Sensors: Joint angle position, leg load

Frame composition: Aluminum

Batteries: None

External power: Compressed air

KLOC: About 30

Hours to develop software: About 4,000

Funding agency: Office of Naval Research

Project status: Ongoing

Information from: Roger D. Quinn

The product of a long quest, Robot III, an artificial cockroach built by mechanical engineer Roger Quinn (left, in blue shirt) and biologist Roy Ritzmann at Case Western Reserve University in Cleveland, required seven years to construct. (Quinn directs the Biorobotics Lab at the university.) Based on the death-head cockroach, the artificial bug is made from pneumatic tubing and high-grade aluminum alloy. Although it cannot yet walk—the robot can only lift itself upright from a crouch—Quinn and Ritzmann hope eventually to have it scurrying up slopes and avoiding obstacles, all the while carrying a payload on its back.

In a Kafkaesque scenario, an anesthetized female cockroach is pinned on its back in a petri dish coated with a rubbery goo (above). Guiding himself by peering through a microscope, James T. Watson, a staff researcher in Roy Ritzmann's lab at Case Western Reserve University, inserts the wires from thin pink electrodes into one of the insect's leg muscles. The electrodes will be used to take measurements of the insect's leg muscles when it moves—information that will be used by roboticist Roger Quinn in his roach-robot projects (previous pages).

animal in terms of locomotion. That means walking, running, turning, climbing—all the things that the cockroach does really well. The way we are doing this, we call biology by default. If we know how the biology solves the problem, then that's how we solve the problem on this robot. If we don't know, then of course we just use whatever will work. But if the cockroach does it, then that's the way we want to do it here. Because we want to understand the biology as well as build a good robot. We hope we'll reap rewards that otherwise we wouldn't. The animal does it that way for a reason.

You and your colleagues have been working on these bio-robots for quite a while.

Quinn: Yeah, the three of us have been in this since 1990, something like that. Randy Beer actually started the whole program off with his dissertation before he became a faculty member at Case Western. He was doing computer science work in biology—not robotics—it was a simple simulation of a roach-like robot. And that's why we built Robot I—to verify his simulation work. He wanted to prove it out in an actual robot.

And now there's a Robot II and a Robot III.

Quinn: Right, and Robot IV—we've built a prototype leg for Robot IV already. A couple of them, actu-

ally. One of the things that's wrong with Robot III— as you build them you see what's wrong—when you look at its leg, you have to look really hard to see the segments. What you see are actuators and hoses. That means that the scaling of mass is not right.

The pneumatic actuators are too big?

Quinn: That's right.

Is that a significant problem?

Quinn: I don't know if it's a problem or not, but it means that we can't do the things that Bob Full talks about. What Bob Full likes to point out—and he's absolutely right—is that a lot of the things that the animal does, it does through the passive properties of the muscle. Essentially, the nervous system just tunes the stiffness of the muscles. So the nervous system may just be doing fine-tuning in typical walking and running. What goes wrong here is we've got all the masses all messed up. It's not really bad, but it's not as good as it should be, so that we could get the properties tuned properly.

If you had a different kind of artificial muscle instead of these pneumatic, pressurized-air actuators, could you more closely approximate the animal?

Quinn: Yep. Something we would really like to have is artificial muscle. The Stanford Research Institute's [SRI] artificial-muscle program is leading-edge in

terms of actuators, and we want to build robots. In a year or two I'm hoping that we can use the stuff they're developing now, because we need something that we can put on the robot and make the joint move. And put it on all the legs and make the robot go.

The SRI muscles use polymers that change shape when an electric charge is applied?

Quinn: I love the SRI actuators because there's no compressor; compressors tend to be inefficient. You do have high voltage to deal with and you have to be careful about shielding the electronics because the voltage can kill the control system, but all that can be done. It would be nice to go from batteries to electric current, which drives the actuators, because it's a one-step process.

This robot will look a lot different at that point.

Quinn: Oh, yeah. Somewhere we have some photos [taken] before we put on the actuators. It looked so much nicer. It looked much more like the animal because you could actually see the segments. If you stoop down you can actually see them, but when you're standing up, you're seeing actuators. And the trouble with pneumatics is that it's hard to control. If you look at industrial robots, the only time you ever see pneumatics is when you

have a mechanical stop here, a mechanical stop there, and no in-between stop. When you try to control it in some other position, it's hard because a pneumatic system is so slow to respond.

It they're so problematic, why are you using them?

Quinn: We used motors for Robot II. Motors are a lot easier to control. We won an award with Robot II for its fantastic insect-like walking, but it could not go up a slope because it didn't have enough power. It couldn't climb over much of anything. If we tried to make it autonomous, it would be slower and without much power. So you're kind of stuck if you're using motors.

So pneumatic actuators were the best choice?

Quinn: We wanted something that had lots of power for low weight and that's how pneumatics came in. This robot can actually do pushups with a 30-lb. [15 kg] payload. Well the robot weighs 30 lbs. so it's essentially doing marine push-ups. It's actually capable of jumping. It has jumped. Not on purpose, but it has jumped [laughs].

Did it get hurt?

Quinn: No, but it's given Gabe Nelson, the graduate student who's doing the control work, a few bruises now and then.

I think you might want to find something that won't start breaking your graduate students [laughter].

Quinn: We're looking.

Case Western research biologist James Watson (above) nudges a cockroach onto an insect-sized treadmill, intending to measure the actions of its leg muscles with minute electrodes. To ensure that the roach runs on its course—necessary for his research into animal locomotion—Watson coaxes it onward with a pair of big tweezers. In the experiment, the electrode readings from the insect's leg are matched to its movements, recorded by a high-speed video camera. Reflecting on his collaboration with engineering professor Roger Quinn, who is using Watson's data to build robot cockroaches, Watson says, "The central problem of collaborations between scientists and engineers is that engineers are supposed to produce something that works. Failure is not an option. Whereas we scientists are interested in understanding things. If it fails, that tells us something. As a matter of fact, it's much more interesting if it doesn't work, because that tells us what we don't know."

Good-naturedly donning fishy swim goggles for the camera, Yuuzi Terada, an engineer at Mitsubishi Heavy Industries, stands at company headquarters with a pair of the sleek robot fish he constructs. Gray's Paradox asks the question why fish, with their slim muscles and small fins, can accelerate so quickly. Researchers have long hoped that unraveling Gray's Paradox will allow them to build safer, faster nautical propulsion systems. The dream is shared by Terada and other researchers at Mitsubishi, who have long thought that fish fins might serve as a model for a new kind of propeller that would make underwater vehicles faster, more stable, and more maneuverable. For now, though, the company principally hopes to market Terada's battery-operated fish to aquariums and museums around the world. One idea is to build models of unique fish, such as the coelacanth (the larger of the two fish—the other is a modern sea bream). The coelacanth was known so well from fossils that until the 1970s scientists thought it must be extinct. Earlier sightings of this big, strange fish in the Indian Ocean gave rise to legends of sea monsters.

Working for Scale

"Avoid the wall, avoid the wall," John Kumph says, paying out a thin power cable between two fingers. On the far end of the cable is a robotic fish called, inevitably, Wanda. The silicone rubber-sheathed fish, Kumph's doctoral thesis project, is swimming underwater in a large tank at the Massachusetts Institute of Technology. "There you go," Kumph says. "Good fish."

The fish's tail flops smoothly back and forth in the water, begins a turn—and hits the wall. "Oops," Kumph says. At this point, it seems, Wanda is still a swimmer, not a sensor. Leaning over the tank, Kumph grabs for the robot's nose and pulls it out of the water. "Whoa. You know, I think she's sinking. She's taking on a bit of water—not too much, though."

Scientists hope that Wanda—and other robot-fish projects, such the 150-kg robot tuna at Draper Laboratories in Cambridge—will resolve Gray's Paradox, according to which, fish should simply not be able to use their slim muscles and skinny fins to move themselves so quickly through the water. (Some scientists don't believe that Gray's Paradox exists, and Kumph is one of them.) Researchers obtain the funds to pursue this apparently arcane

question because the Pentagon believes that robotic fish might have another use: detecting sea mines. Kumph's own goal at this point is graduation with his sanity intact (he has since done so). "I've been working on this fish for four years," he says. "It's time." And it's prime time for Wanda. The fish has been added to the stable of robots at the robotics company iRobot.

Faith: What do people ask you about Wanda when they see it in the tank?
John Kumph: Everyone asks me, "What is it supposed to do?" I tell them, "Swim" [laughter].

And when people who know about robotics look at it, what do they say?
The first thing that they say is, "Wow, you do it in the water!" [Laughs]. The water thing actually turns out to be tricky, because there are electronics in there and they don't like to get wet. And then people like the flexibility. Of course, there are no fancy motors. I would like to do fancy motors sometimes.
This has been a worthwhile project for you?
Definitely. I have learned so much about how to make things, about robots in general, about electronics, mechanics, and fish. It is amazing how hard it is to copy nature. We are so far away from being able to do it.
Robotics is really in its infancy.
A lot of times when you read the newspaper, you get this sort of modernist perspective that human beings are almost to the top. We know so much, we can do all these magic things. But the truth of the matter is, we have so far to go. Even when it comes to technology, we are nowhere. I can't even get this thing to work, half the time. Well, I can get it to work more than half the time now, but at first it was definitely less than half the time. It's amazing. What it's supposed to do is so simple.
How does Wanda swim?
By wiggling its body. The first thing was figuring how to get it to wiggle its body right. And so I studied fish and the shapes they make in the water. And then I built this machine that tries to make the same shapes.
How did you do it?
You take still pictures of fish swimming—a lot of people did that as early as a hundred years ago. So I studied those, and then more recent pictures of the same kind of thing. And then I built a machine that could make those shapes. I worked from the outside in, because I wanted to copy the outside shape of the fish, as well as the way it could bend. It is kind of like a vacuum hose, like a spiral spring, and then there is this plastic [covering the coils of the spring]. Then there is a machine on the inside of it that bends the hose [in a fish-like, undulating motion], and that is how this actually works. There are three motors in here, and one for each fin. The idea for these fins is that at some point I was going to get a little forward speed and start diving. But I haven't gotten there yet.
Will you get there?
Maybe, maybe not. I am happy that it swims at all. The really interesting stuff is not just swimming forward, but the turns—that is something people hadn't done before. Turning, it turns out, is really tricky. There is a lot of timing involved, and getting the shapes right, and no one is really sure how it is done.
When I look at this fish swimming, I think it looks pretty much like a fish. When you see it, what do you think?
Not a bad first try. I have gotten to the point where it's does a decent turn. It is not as tight a turn as I would like, but it's not bad.

Robo Specs

Name: Robopike or Wanda

Origin of name: Robotic pike and movie title, respectively.

Purpose: To develop new forms of underwater propulsion.

Creative inspiration: Watching fish fly around underwater, *Blade Runner.*

Height: 15 cm

Length: 81 cm

Weight: 3 kg dry

Actuator type, number, and kind: Futaba 9303 servos, Tower Hobbies TS-51

Vision: None

Sensors: None

Frame composition: Fiberglass

Batteries: NiCd

External power: 17 V

Project status: Ongoing

Information from: John Muir Kumph

In the water, pike can accelerate at a rate of eight to twelve g's—as fast as a NASA rocket. To scientists, the speed is inexplicable. In an attempt to understand how the flap of a thin fish tail can push a fish faster than any propeller, John Kumph (left), then an MIT graduate student, built a robotic version of a chain-pickerel—a species of pike— with a spring-wound fiberglass exoskeleton and a skin made of silicone rubber. Now under further development by iRobot, an MIT-linked company just outside Boston, the robo-fish can't yet swim nearly as fast as a real pike, suggesting how much remains to be learned. Kumph hopes that some quality of the skin will be the key to learning the secrets of fish motion. If the speed is due to the undulation of the fish, he says, the force will be much harder to harness in a ship.

Claw and Order

Robo Specs

Name: BUR-001

Origin of name: Biomimetic Underwater Robot 001

Purpose: Because it isn't there....

Creative inspiration: Teaching a neurophysiology/behavior course where the students design an animal.

Height: 20 cm (25 cm when walking)

Length: 61 cm

Weight: 2.9 kg

Vision: None

Sensors: Compass, tilt sensors; others under consideration

Frame composition: Plexiglass and machined Delrin

Batteries: Yes

External power: Wall socket

KLOC: I haven't had time to count them up.

Hours to develop software: You really must be kidding....

Cost: Cheap ... how many do you want? However, consider the answer to the previous question. Those are my hours.

Project status (Ongoing or complete): How would Wilbur Wright have answered this?

Information from: Joseph Ayers

Joseph Ayers (right), head of Northeastern University's Marine Research Laboratory, has been researching lobster locomotion for more than twenty years—looking at the many different movements of a lobster's legs as it climbs and walks forward, sideways, and backwards. Based on Ayers's studies, staff researcher Jan Witting is building a robotic lobster that will capture in detail the behavior of a real lobster. Unlike many biomimetic researchers, Ayers is not so much trying to abstract the principles of a living creature as come close to duplicating them mechanically. On the other hand, the project has enough potential for sweeping mines that it is funded by the Defense Advanced Research Projects Agency. Although its eight legs are in place and working, the battery-operated machine still lacks the tail and claws that will stabilize it underwater—real lobsters are buoyant, and almost float along the ocean floor with their claws and tail extended in a triangular shape that maximizes their stability.

Joe Ayers and Peter are deconstructing lobsters at dinnertime on Ayers's kitchen counter. The discussion hopscotches between the merits of the lobster as a model for an underwater de-mining robot and the proper technique for claw meat extraction. "There's plenty here for you to practice on," bellows Ayers to Peter. Prudently, Ayers's wife, Nancy, and I are staying well out of the way.

Ayers, the director of the Marine Science Center at Northeastern University, and his military-funded team of researchers are working on the premise that the methods lobsters and lampreys use to find and identify prey are exactly what autonomous underwater robots need to use to find mines. Just as Ayers, the author of a self-published fusion cookbook, knows what he wants in a lobster dinner, he knows what he wants in a lobster robot.

Faith: Some biologists who work with roboticists want to accurately reproduce every part of an animal. I take it that is not your method?

Joseph Ayers: To me, the big picture is the major organizational or governing rules by which you operate the thing. Others are very focused on their robot behaving exactly the same way as the animal. I want to say which details are important, eliminate the non-essential details, and focus on the governing principles.

Well, what's important in the lobsters you study?

Their ability to go and find something on the bottom of the ocean is unparalleled. If they want to go out and find trap bait, they can get through anything. It's that kind of performance capability we want to capture. And energetically, they are very efficient. We are able to so closely mimic the way the animals operate in the big picture—not the details—that we can really capture these performance advantages.

What are you hoping to achieve?

I think the goal is to build robots that are autonomous. That's the Holy Grail as far as I'm concerned—autonomy. The guys here have built these incredible sensors. They so closely mimic the way the animal's nervous system codes the information that they just fall naturally into this neural-circuit architecture that we're using.

What's a neural circuit for a robot?

I call what I work on neural circuits and I really mean the circuitry that exists within the nervous system of animals. I call what I do biological intelligence.

As opposed to artificial intelligence? AI involves all this software that can learn and evolve—is that important for emulating a living creature like a lobster? Some roboticists think so.

I think AI is good for data mining and all sorts of information-handling processes, but AI as a substrate for robotics? Where is the success? Show me an example of an autonomous robot that can bound adaptively through the forest. Do you know of any that came out of an AI lab? How long have they been at it? Thirty years? You'd think there would be one, wouldn't you?

Not necessarily. Computer science is very young—look how long it has taken chemistry and biology to mature.

I've been to enough of these meetings to see there's a certain level of stagnation.

Well, if there's nothing to AI—

I didn't say that at all. I said AI-based robots. What we're saying is that we have an existence-proof. We have an animal. Animals have evolved to occupy any niche where we would ever want to operate a robot. So for that niche in the environment, there should be an animal robot that will do what you want it to

do. We now know that the nervous system of animals is conservative. The same basic organizational units of the nervous system are involved in the motor systems of all animals. If you know how to use that architecture, you should be able to build a robot to operate in any environment on the planet. *So you can simply program the robot to do what it's supposed to do—you don't need all this fancy superstructure of adaptation and learning software.*

The reinforcement-learning business is what is in vogue in computer science. Many computer scientists think that understanding the brain is a trivial problem compared to understanding computing. So they are all hung up on psychological concepts. They grav-

itate toward psychological approaches, and of course learning is very big in psychology. Well, you won't see many courses on the psychology of lobsters.

There's a reason for that.

Exactly. They are just much simpler animals. They've evolved to a much simpler behavior set. People are flabbergasted that you could blind a lobster and it gets along just fine. That shouldn't be a surprise to anybody. People are so hung up with visual interactions with the environment. You take them away and for us that would be paralyzing. For a lobster, it's a non issue.

Then why does the lobster have eyes?

I think they are used largely in prey-capture situations.

Peter's Notes

I photographed Joe Ayers with his lobsters, real and robotic, while deep inside a concrete bunker beneath a seaside cliff near Boston. It seemed dramatically apropos that his research was housed in a World War II naval bunker—the Defense Advanced Research Projects Agency and the Office of Naval Research were spending millions on it. Afterward, we traded books we had written. Joe's was a cookbook. On our next visit he prepared lobster for us at his house.

Knight Light

The first time Frank Kirchner demonstrated Sir Arthur to marine biologist Joseph Ayers (previous pages), one of the robot's legs fell off. But the demonstration was anything but a disaster. "It still walked," he says, which proved its adaptability. Kirchner, an artificial-intelligence researcher at the German National Research Center (GMD), designed the six-legged machine to search for and approach brightly lighted areas. Sir Arthur's legs regularly fell off, both Ayers and Kirchner recount with humor, but the results were nonetheless impressive enough that the two men are now collaborating on a scorpion robot. Funded by the Pentagon, Kirchner's scorpion robot and Ayers's lobster robot will hunt for and identify land mines.

Faith: How is the scorpion robot going?
Frank Kirchner: I don't have any real robot to show off yet, but I do have our first leg prototype. And we're working on a second leg design, which will use much bigger motors.
Joe's lobster robot is being built to find underwater mines. What role will your robot play?
The original idea was that the lobster [Ayers's robot, previous pages] would find mines under water and the scorpion would find mines on land. In this DARPA [U.S. Defense Advanced Research Projects Agency] project we want to build the basic platform, where the scorpion will be able to navigate and maneuver around. Then we can implement things like fancy sensors that people come up with.
Why are you doing the robots together?
What we want to do with both is join the control architecture, so we are both doing the same reverse-engineering of the animals. Basically, having models of them that we then implement on the robots.
Given your background in artificial intelligence, I would imagine that you're using an AI approach.
I studied computer science and neurobiology. I got my degree in both, so that was my whole idea in the beginning—to understand intelligence.
So do you understand it yet?
No, not yet [laughter]. Basically, Joe and I are working together doing the reverse-engineering of the animals. We look at them behaving in the tank. How do they overcome obstacles? How do they navigate in a maze? And then we try to puzzle out small elementary behaviors. This is where I can come into play, using machine-learning algorithms to sequence those behaviors to form complex behavioral acts that have some meaning in the world.
What kind of learning?
Reinforcement learning. It's like training your dog.
Like rewarding a dog for sitting when you tell it to sit.

But dogs are pretty smart—what is the level of intelligence you are talking about here?
We do want to build these machines to be able to learn from their interaction with the environment and to improve their intelligence over time. Right now we are at the low-level, behavior-based approach, where you deal basically with motor intelligence and kinematic intelligence. "How do I walk?" "How do I stretch my extremities?" "How do I eat?"
Your collaborator, Joe Ayers, says there isn't any learning going on in lobsters at all, and the software for lobster robots doesn't need to be able to learn, either.
It's very true that Joe would argue with this. It's not by chance that we got the DARPA grant after Joe got his grant. I think that the funders see the beauty in both approaches and in some areas Joe and I are completely identical—in the idea that you can come up with a robust set of low-level elementary behaviors, where the creature runs into a situation and understands it with its sensors and has a fixed reaction to whatever the situation is. But on top of that, I would see a layer of adaptive-learning behaviors that are able to modify the low-level behaviors in order to appropriately react against changes in the environment.
But for now, you're building a machine that has programmed reactions?
We're building up this robust base layer of behavior that lets our system be reactive. It's not bumping into stuff, it's finding odor sources and all sorts of things. But then, I would say that you need another layer on top of this. A simple example—you walk on this smooth ground and then you walk into grass. Now you have totally different traction, so you want to adapt your walking pattern. If you have a fixed mode of walking, then you are not very efficient. And efficiency is important in the animal kingdom. I don't know if Joe would disagree that mammals have this layer of learning power on top. He would certainly—and he may actually be right—disagree on whether that layer is present in invertebrates.
What are your goals?
We want to have solar cells in it so it can actually go out for days. When it feels that the batteries are empty, it shuts everything down and reloads. When it's loaded up again with solar energy, it continues on its mission. We want to build a robot that's able to survive in the real world for longer than its battery life. Using this biomimetic approach, we're trying to build robots that can survive in the environment for as long as they can recharge their batteries. That can optimize their behavior in that world by learning from it and memorizing its interactions.
Staying active as long as it can get energy? Sounds like life to me.

Robo Specs

Name: Sir Arthur

Origin of name: The mystifying feeling you get when you see the work of so many months actually walk along the corridor.

Purpose: A proof of a theoretical concept in machine learning.

Creative inspiration: The challenge to create an autonomous creature that learns on its own.

Height: 10-18 cm

Weight: 1.8 kg (with batteries)

Length: 28 cm

Vision: None

Sensors: Touch, ultrasound

Frame composition: Carbon fiber and aluminum

Batteries (type, duration): 6 V / 2 A, 30 mins

External power: Optional

KLOC: Only about 150.

Cost: $10,000

Project status: Ongoing

Information from: Frank Kirchner

Photographed at a baptismal font in the chapel of Schloss Burlinghoven, a nineteenth-century castle on the campus of the German National Center for Information Technology, the walking robot Sir Arthur (left) stands with its creator, research scientist Frank Kirchner. Sir Arthur began as a relatively simple robot with sonarlike "vision" that prevented it from trapping itself in corners and snagging itself on obstacles. It was successful enough that Kirchner obtained funding from the U.S. Defense Advanced Research Projects Agency to assemble a team of researchers from diverse disciplines—computer science, math, physics, and electronic and mechanical engineering—to build an enhanced, solar-powered version that can cross rough outdoor terrain. He is also building a scorpion robot and collaborating with Joseph Ayers (previous pages) on his lobster robot.

Name: Troody

Origin of name: Late Cretaceous dinosaur *Troodon*

Purpose: Museum dinosaur or early-life robots for exhibits

Creative inspiration: From looking at pictures of dinosaurs in dinosaur books.

Height: 46 cm

Length: 122 cm

Weight: 4,5 kg

Vision: None

Sensors: Joint angle/torque and pitch/roll

Frame composition: Aluminum

Batteries (type, duration): NiCd, 30 mins

External power: 24 V for tethered operation

KLOC: 1

Cost: $4,000

Project status: Ongoing—Troody can take several steps in place with fair reliability as of this writing.

Information from: Peter Dilworth

Peter's Notes

I have photographed lots of research projects that I knew would never work as planned, including Reagan's Star Wars weapons research at Los Alamos, Biosphere II in the Arizona desert, a flying car in California, a nuclear-resonance battery in Oregon; and human blood substitute in a baboon experiment. A lot of very smart people spent a lot of time and money trying to make these things work. It never happened. Fortunately no one died in these projects, except the baboon. Peter Dilworth's walking robotic dinosaur seemed like another candidate for this category—he was always working, but the robot wasn't. But eventually I saw it stand up on its own power. Because I know Dilworth has the perseverance of a pit bull with an engineering degree, I'm sure he'll get it to walk around his lab sometime soon. But I am afraid his dream of two-legged robot dinosaurs wandering around in museum halls is a long way off.

Jurassic Classic

Peter Dilworth powers on his four-year-old robot Troody, which he has modeled after a birdlike dinosaur of the genus *Troodon*. When the robot has warmed up, it will stand, promises Dilworth, a staff researcher in the Massachusetts Institute of Technology's Leg Laboratory. "It can't walk yet," he says, "but it can stand." At this point, the robot is hunched down like a nesting bird.

"Okay, let's see what happens," murmurs Dilworth. He taps a command on the keyboard that controls the robot's onboard computers. In response, the robot dinosaur pulls itself erect. "It's slow now," Dilworth says. "It's doing slow, static standing. I'm going to speed it up a bit."

Suddenly the laboratory is filled with the spastic, staccato thumping of the robot's feet against the desk. The nesting goose is now doing Darryl Hannah's death scene in *Blade Runner* and Dilworth grabs the robot by the neck before it can injure itself.

"It wasn't supposed to do that," he says. "But I know why it did. The problem is the infrared sensor—the noise from the lights. There's a probability that once every two or three days it will get a false command from the lights, which is really annoying. I have to fix that."

"That's what happens to me when I have too much coffee," remarks Peter, the helpful photographer. Dilworth is persistent. "Let me try again," he says. "It's a testament to the robustness of the hardware that it can spaz out like that and still work." He sets the robot down on the desktop and tries again. Again the robot flops wildly.

"Nope, that wasn't the problem. I've left some variable in the code with an invalid value. I've got to find out what it is. This is called being in demo mode. It hasn't done that in months."

Demo mode, as we have come to learn, is engineering jargon for the malfunction of a device that has repeatedly worked well for you but ceases to operate when you attempt to demonstrate its capabilities to others. A well-known phenomenon, it is most often experienced before funding agencies, university presidents, academic conferees, and PhD review boards. Luckily for Peter Dilworth, his demo mode is happening only in front of us—although of course we are putting it into a book. Dilworth is good-natured about it, though. I tell him that I'm sure that Shigeo Hirose (see pages 89 and 192), the robotics maven at the Tokyo Institute of Technology, has a demo-mode day now and then.

Peter Dilworth: I admire Hirose a lot. When we went to Japan to visit his lab, they had this cluttered room with junk in it and they'd say, "Oh, let us show you this robot we made two years ago." And they'd pull

some junk over and then there'd be this robot under some stuff and they'd turn it on and it would still turn on! It was like a scene from *Star Wars*. And just the fact that it worked and was ready to go—that made a huge impression on me. Hirose is God.

Faith: There are a lot of people who think that. Your biggest challenge is that you designed Troody to be autonomous.

Yes, that doubled the amount of time it took to build the robot.

Doubled, really?

Yes, because it had to be very lightweight.

And you had to put all this control hardware in for it to be autonomous. How far along would you be if it weren't all onboard?

I would have had another three pounds of weight to work with. So, because I had to shave off three pounds of weight [for the motors to be able to make the robot walk], I had to make parts super lightweight. It's kind of like aerospace, basically. It takes time to optimize everything.

It's like birds, with their hollow bones.

Exactly. And battery technology is very poor, so Troody's batteries will let the robot walk for only twenty minutes to a half an hour, which really isn't very long, but that's the best you can do with modern batteries. So anybody who builds an autonomous robot impresses me because that means they really had to think about a lot of stuff. They couldn't assume that there's this big power station with a wire coming in.

You just updated Troody?

I gave it a double hip replacement because the hip joints were too weak before and they were getting loose. And the instrumentation that measures the joint angles wasn't working that well. I redid that completely to make it work better. That stuff wasn't really important to make the robot stand up, but for stepping it's really important to have it working well.

And is it stepping well?

Yes, but I'm trying to get the robot to go beyond stepping and begin to do more walking and bigger steps, that kind of thing, so I need to make sure the hardware is as good as it can be before adding a complicated control system.

What is the control system that you're graduating to?

The previous control system is a very simple few if-then statements. But the new one is going to be different. It's going to be either an adaptive controller, maybe even a neural network, or at the very least some kind of a large tabular thing where, instead of having a simple line of code, you have a huge array of numbers that it indexes into, depending on what the joint angles are.

You've been working on the hardware for a really long time. It must feel great to finally be close to finishing.

It's really exciting. It's definitely to the point where

it's time to say, "I'm done with hardware and it's time to dive into the control system." That was the hard part. The fun part is the control system, in my opinion. The hardware is interesting but it's not where the mental challenge is.

There's no mental challenge in getting the hardware right?

Oh yeah, but the control system is more procedural. Hardware is more just three-dimensional, spatial. You just use a different part of your brain, I guess.

And do you have a plan for Troody in the future?

The robot is going to be in a museum near you in a few years. The goal of the robot is to walk around and look like a dinosaur. When I get this prototype working, hopefully in the next year or so, then I'm going to be looking for venture capital.

Imagine I'm a venture capitalist and I tell you, "Okay, it walks like a dino, but the mechanism inside is insanely complicated." What do you tell me?

Well, the leg hardware has been through at least three thousand sit-stand cycles. Which is one of the hardest things that a robot will ever do. And I've only had to

replace one gear so far, so the hardware is extremely robust. The other side of the robustness is a good control system, to keep the robot from destroying itself.

From twitching and destroying itself?

Yeah. As you see, I haven't gotten to that point yet. If I can, I can. And if I can't, I can't. It remains to be seen. It's important to note that this is a lightweight design. It has to carry its own power and everything onboard. Just the concept that you can build something that's lightweight but robust enough to go through repeated cycles—that it is feasible—that's the point I'd try to get across to a venture capitalist. Building a robot like this is a way to get the technology out into the world. If you build a little lab robot that is useful to no one, that may be an interesting thesis project, but that robot is not going to do anything for the world. I'm trying to create an application of this technology that could have an immediate benefit. It doesn't need to be super-reliable. It just needs to work in a museum for a few minutes. And once it's out there, I hope it will create its own little niche.

Designed to run freely on its own, untethered to any external wires, Troody is modeled after *Troodon formosus*, a carniverous dinosaur that was reputedly the most intelligent creature in the late Cretaceous era. Troody's designer, MIT Leg Lab staff researcher Peter Dilworth (above), wants to put such prehistoric creatures in museums. In his scenario, viewers will be able to control them as they roam the museum hallways—interactive displays that will convey something of what it must have been like when these creatures ruled the earth. The idea has been embraced by Gregory S. Paul, a prominent paleontologist. Alas, Dilworth has some distance to go: the robot now can only stand up, sit down, and take single steps. Even that is not always possible; just before this portrait was taken at the university boat dock on the Charles River, Troody lapsed into immobility.

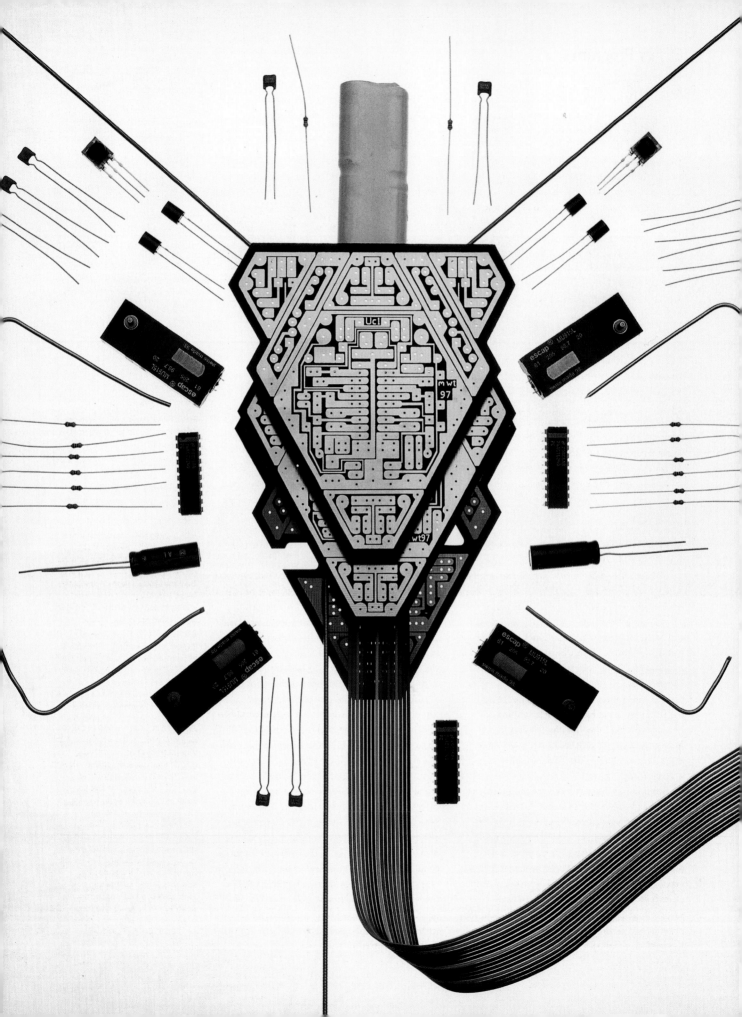

Buggy Program

"This is my main lab area," Mark Tilden tells us about the dining room in his home. On a dusty glass-topped table are several multilegged, flat-bodied robot creatures the size of television remote controls. They have whimsical names like Ogre, Snakebot, Walkman, and Strider, and scientific designations like CTD 1.4, VBUG 3.3, and GPIM 2.4.

On a workbench in an upstairs bedroom, more small robots are under development, scattered amid snippets of wire, solder, and sheets of specially designed wafer-thin circuit board. Still more are tucked into cases that shuttle back and forth among exhibitions and museums. All the machines are small and all are analog controlled—there's not a single digital processor among them.

Officially, Mark Tilden is a robotics physicist at Los Alamos National Laboratory in New Mexico, but he is principally interested in what he calls "biomorphic" robotics—the work he does on these tiny machines. When I ask if Los Alamos has claims on them, he says that he owns the biomorphic work "patent, copyright, and doo-da."

Tilden's robots are made from transistors, motors, sensors, and anything else he cares to throw in—many of the parts salvaged from cast-off cameras and videocassette recorders. He endows his machines with what he calls "nervous networks," which function much like the neurons in the nervous systems of animals.

Unlike conventional robots, which are usually based on digital technology, Tilden's nervous network machines are based on analog technology—continuously changing waves, rather than ones and zeroes. Rather than writing elaborate programs to make his robots walk, he designs circuits that automatically seek a desired state—and motion is usually the result.

When his robots encounter obstacles, they frantically try to get back on track, sending out a blizzard of almost random impulses until they manage to scramble past the obstruction. The results are startling: Tilden's devices can skitter over almost any terrain.

Given the contrast between the fluid motions of his inexpensive machines and the stiff, clumsy behaviors of many more costly robots, it is little wonder that the bombastic Tilden enjoys being a vocal skeptic about the digital world, and even professes to enjoy the drubbing he sometimes gets from those in that world. "I don't use computers except to comb the nits out of my beard," he says expansively, as I settle into the chair next to his laptop.

He's exaggerating—Tilden returns our e-mail with unnerving speed. At the same time, he's serious about advocating analog. Because analog technology is often regarded as clumsy and outdated, I ask, "Do you use your philosophy about analog robotics at work?"

"Yes," Tilden says, "for proof-of-concept prototyping, and then later with processors for compatibility. What I've found out is you can only be a rebel for so long. The quick turnaround analog provides can run on digital controllers, and it's important that other people are able to understand what you do so they can run it too." However, retrofitting his ideas for the digital world is uninteresting to Tilden, and he's got the latitude from some big government grants to continue his idiosyncratic research into small, analog, biomorphic robots.

Faith: What's wrong with taking the digital approach to robotics?

Mark Tilden: Nothing, provided industrial puppet robots are all that is required. However, as soon as a device has to interact with complex worlds, like a forest, or a field, or a human, then digital hits the complexity barrier, where a small behavioral improvement can cost thousands of man-hours. We're already seeing this with computers, and they just have to sit on a desk. If laptops did have to move and work in the real world, the complexity required would continually require precise, predictive updates. This just isn't easy to do in the language of digital logic. They're too fragile to take risks, like living things.

Even worse, it may not be necessary. Nature uses analog answers but we can't build those. I believe there's another way, ideal for robots themselves. That's why I'm exploring concurrent-signal analog devices, to see if there might be other solutions to making a machine think it's alive.

Are all conventional robots puppets, as you call them?
Almost all. You should always watch out for what I call Wizard of Oz demonstrations. "Pay no attention to that graduate student behind the curtain—I am the great and powerful roboticist of Oz." If you see a machine with a whacking great big cable or antenna coming off of it going to a supercomputer run by various grad students, then you are looking at a Wizard of Oz demonstration. If it repeats a behavior pattern, then you've got a puppeteer on tape. There are over 10,000 special-effects masters on the planet right now building wonderful things for Spielberg, et al. But if you count the number of people who are really researching robots—who are trying to go beyond the fiction—the total is small, compared to the people who are making Jim Henson machines that have to run using human control.

Who do you respect most in the field?

Robo Specs

Name: Unibug 3.2

Origin of name: Contraction of Unicore Bug

Purpose: Research into adaptive minimal control for autonomous agents; experimental high-speed walking platform to test 20-transistor Unicore controller.

Generation: 3.2—third of genus.

Height: 5 cm

Length: 24 cm

Weight: 206 g

Vision: Active modulated emissive detectors and passive ambient IR to UV, low-resolution detection

Sensors: 4 motor-load, 3 tactile

Frame composition: Carbon steel, copper buswire, piano wire, silver solder, FR4 PCB substrate

Batteries: 4.8 V 80 mAh NiCd (rechargeable)

External power: None

Person-hours to develop software: "Convergence" time approximately 10 minutes.

Cost: Motor $170, batteries $4, electronics $2, sundry $40

Project status: Complete

Information from: Mark Tilden

———————————————————

Like a dissected mechanical insect, the hand-sized walking robot Unibug 3.2 (left) reveals its fifty-component construction to the camera's gaze. Designed by Los Alamos researcher Mark Tilden, Unibug uses simple analog circuits—not the digital electronics that are in most robots—to poke its way around an amazing variety of obstacles. Digital machines must be programmed to account for every variation in their environment, Tilden argues, whereas analog machines can minimally compensate for new and different conditions. When the machine is set down, the stress disturbs the waves in the circuitry by which it moves. Unibug tries to compensate, continuing until it finds new movements that allow it to walk with the least possible energy. In changing surface conditions—walking on sand, for example—this process makes the machine look as if it is lifting its legs higher and higher until it is doing a sort of Rockettes-style kick to work its way over the surface. But the odd, low-tech procedure works. The machine strides confidently as long as its batteries last—unlike many of its bigger, more expensive robot brethren, which often can barely put one foot in front of another.

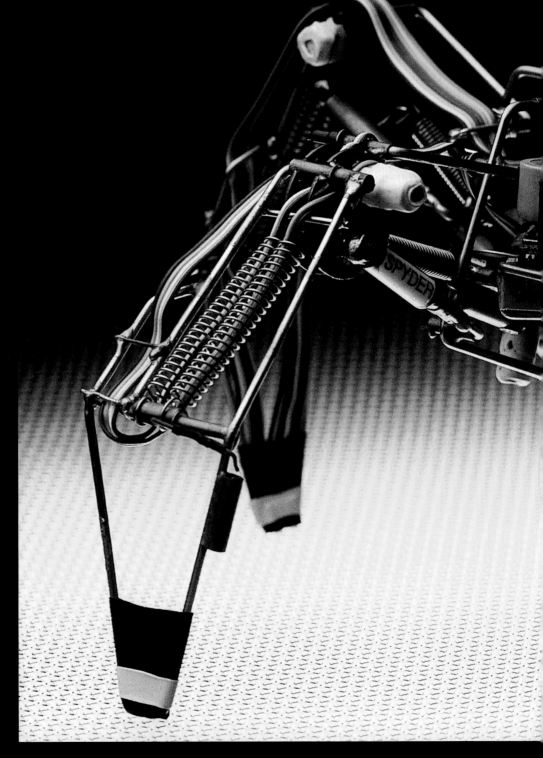

As Mark Tilden's Spyder 1.0 approaches like a tiny but menacing arachnid, its circuits try to optimize actions—walking in this case—with minimal energy. Perturbed by the environment, its patented "nervous net" seeks the minimum state, its legs moving almost randomly until it succeeds—at which point the circuitry seizes on a new minimum and maintains it. In 1990, Spyder 1.0 was the first walking robot to use Tilden's nervous net control system. When Tilden first achieved such complex behavior from such minimal components, the results astonished some roboticists. His creations, one journalist noted, were easier to build than they were to explain.

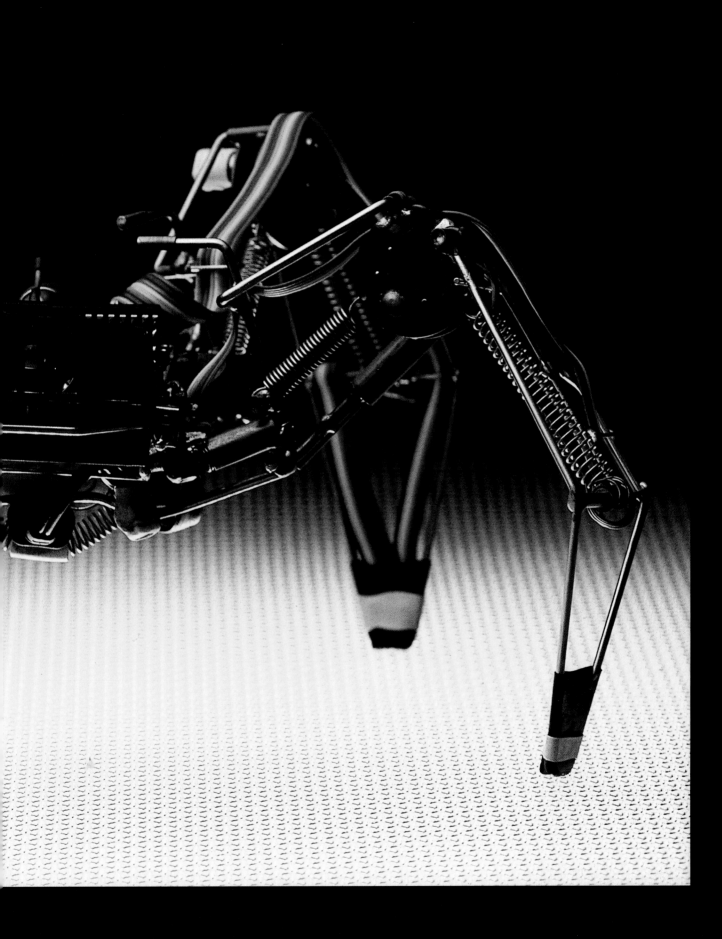

"Nothing in nature is digital," says researcher Mark Tilden, who created Unibug 3.1 (below). "Everything's analog—and analog can do better." Unibug 3.1—a slight variation on the disassembled model pictured on page 116—is an example of what he means. Although built of simple, off-the-shelf components, it can walk easily on a remarkable variety of surfaces, striding from a film of shallow water into deep sand without stumbling.

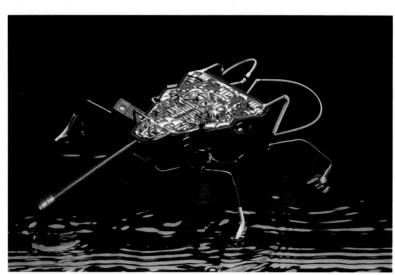

Robo Specs

Name: Spyder (see page 118)

Origin of name: Nickname given by visitor—official name is VBUG 1.0.

Creative inspiration: Horribly over-ambitious vehicle to act as platform for nervous-net controls research. Worked regardless. First functional Nv biomech, Dec. 3, 1991. Crashed into concrete pillar same day. Resurrected Jan. 1994.

Height: 12 cm

Length: 24 cm

Weight: 720 g

Vision: 44 kHz IR emitter-detector (from Pentax portable camera)

Computer: None

Cost: About 50¢ in new resistors

Project status: Still operational after 9 years.

Information from: Mark Tilden

Stephen Jacobsen [see page 216]. Without a doubt he's the roboticist's roboticist. He's not just done the science, the engineering, and the business—he has pulled it all together and tried it 25,000 ways from Sunday and packaged it out into the real world. We all live in his shadow, just hoping that some day he might actually find that hit that science fiction says is out there but reality says is not [laughs].

What specifically intrigues you about him?

It's that he did the entire package. It's very easy for a roboticist to build a bug and never do anything else. Or just sit in a university and teach courses on design. He didn't do one thing—he did them all. And that's a fairly difficult undertaking in a field that isn't all that defined yet. It amazes me that one person could come up against these incredibly frustrating walls and still do so much in such a broad field.

What kinds of walls?

"One day robots are going to take over—don'tcha know?" Except for some reason we can't even seem to get them out of the lab and over the market wall to get a foothold. The wall is the complexity barrier and other things, but primarily it's that human beings love robots on the screen but will not put up with them in real life. Even when we can make them work as we expect. That's a hard barrier to overcome.

Is that what's wrong with robots—people's acceptance?

The fact is, even where the robots are willing, the human beings are weak. My own personal work—we built de-mining robots, we built house-cleaning robots, we built grass cutters, we built everything. Self-contained, solar-powered, automatic devices that don't require any programming and in many cases don't even require batteries. You just have to put them in the dishwasher every now and again, but they didn't sell! They crawled off the shelves

when we tried to market them, no pun intended.

People have media-enforced expectations of a robot. They don't want robot tools. They demand robot characters. Take a look at all the major artificial characters that you think you know. You'll find out that they're all fundamentally human characterizations.

You mean like the humanoid robots they are building in Japan? Or Cog [see page 58], Rod Brooks's robot at MIT?

No, no, I mean like Robby the Robot from *Forbidden Planet* [the 1956 sci-fi movie version of Shakespeare's *Tempest*], Commander Data from *Star Trek*, and all those wonderful personable robots from *Star Wars*. It's all special effects. That's the real problem and why I keep insisting that a robot is a mythical creature.

The fact is, people want a character that essentially is a metal embodiment of a human personality, except it must be perfect and under your control. You work on something like Cog because you're hoping it will be able to give you an insight into building something that has a level of personality. Rod Brooks found out years ago what I'm just finding out now. Even if you can build a million bug robots for a penny—it doesn't matter.

You have to start building things that fall in with people's expectations. People's expectations of a robot are a faultless Commander Data, wearing an apron, who pushes a broom, and fetches you a beer. Not some sort of machine that roams around the hall or cleans under your couch, even if that's what the robot is best at.

Have you ever built robots using digital circuits?

Conventional robotics? Plenty. But in 1982 I tried to build a digital home robot butler and found that it looks really nice on *The Jetsons*, but it's the last thing you want in your environment. Why? How can you program a robot not to eat the book that you were reading last night? Or to leave your loose laundry alone? Or that toilet paper, even if it is hitting the ground, is not something to be vacuumed up?

The funniest thing was when I walked into the kitchen and there's my robot sucking up the cat's kibble, and the cat is looking at me as if to say, "Get this thing out of my face!" What was awful was finding out that even in a single bachelor apartment I could not come up with all of the subroutines necessary to make this thing useful. Also, I kept tripping over the damn thing.

Trip over your dog or cat in the night, they're soft, it'll heal. You trip over a hard robot, you swear like a trooper and the robot is in for a $5,000 repair the next day. Imagine if your laptop decided to walk around your house. My conclusion is that

getting a robot to make your bed would be like getting a Sherman tank to mow your lawn.

What are we looking at here?

This right here is the famous snakebot. This device has only a twelve-transistor nervous system. I invite you to hold it by the ends.

It's squirming. It feels like it's pushing at my hand but I can push it back.

See, the number of computers that support most other robots is probably astonishing but they can't squirm around like this, or adapt dynamically to loads. Say, for example, we give it a negotiation problem. Now it has to figure out how to get over the obstacle that is in my hand here. Here we go, load once, load twice, try again, get over—. There. Now, why did it do that? Was it following computer algorithms? Was it following written code subroutines? No. It was following the dynamic forces caused by these competent little circuits pulsing back and forth in a reflective way.

When you hold this thing down, or put it on a complex surface, it takes in the world through its motors and modifies that behavior. It might take it a few cycles to actually find an appropriate pattern— there are many—but this thing is one of my most successful creatures. It can pull itself out of buckets without difficulty.

What about this other one that looks like a four-legged spider?

It is a spider. See, when you power my devices on for the very first time, they have no knowledge about what, where, or who they are. They have to learn, in a sense, and they really do. How long does this creature take to learn to walk? Well, if the batteries are with us ... one, two, or three steps. It's got two broken motors but it still pulls itself along.

It's a disabled four-legged spider?

The oldest biomech in my collection. Ten years now and still working. What's interesting about this device is, the operating overhead is really low. I don't have to program anything after turning it on, and it finds its own way to move under duress, damage, and old age. That's a feature, not a bug.

Leaning over the glass-topped workbench in the spare bedroom where he builds most of his robot creatures, Mark Tilden (above) shines a flashlight on what will become the head of Nito 1.0. Many of the components scattered over his desk are simple, cheap, and (by contemporary standards) primitive—many are ripped from junked tape decks, cameras, and VCRs. Nito will be Tilden's most ambitious creation yet. (The name stands for "Neural Implementation of a Torso Organism.") When complete, he says, this easily built machine should interact in a simianlike fashion in its world. Nito's head, mounted on a small circular test pedestal, has sensors that respond to light. Tilden's left hand is tweaking a controller to fine-tune its response characteristics.

Remote possibilities

In the fenced Mars Yard at NASA's Jet Propulsion Labor in Pasadena, California, the remote-sensing robot Rocky navigates a mock-up of the te rain on the Red Planet.

Stellar Objects

Borrowing from *Star Wars*, engineers at NASA's Ames Research Center, just south of San Francisco, are developing a small personal assistant (below) that will hover over an astronaut's shoulder in the space station. Floating weightlessly, the machine could have many uses: patrolling corridors for gas leaks, reminding astronauts about the tasks on their to-do lists, or serving as a communication link when people are busy using both hands.

Spreading its solar-power panels to catch the last feeble light of day, the Rocky 7 (right) patrols the Mars Yard of the NASA Jet Propulsion Laboratory in Pasadena, California. Controlled by an operator (visible in shed window), it is working in dimly lit conditions like those it will face on Mars, which is much farther from the Sun than the Earth is.

Electrical engineer Mike Garrett pulls Sojourner's spare "brain" out of a closet at the Jet Propulsion Laboratory and holds it up for inspection. The circuit board looks a little like the guts of my laptop, but I keep that thought to myself—at NASA, people don't joke (openly at least) about the celebrated robot that in 1997 briefly crawled around on Mars, photographing its surface for an enchanted audience of millions. "Either one could have flown," Garrett says of the spare and its twin, which now rests permanently on the Red Planet with Sojourner itself.

System Engineer Eric Baumgartner shows us the prototype for the next-generation rover sitting in the indoor yard which simulates the surface of Mars. Field Integrated Design and Operations (FIDO) is the space equivalent of a concept car, a demonstration platform kitted out with scientific instruments that will allow a future production version to search for and collect rock and soil samples on Mars.

As Sojourner's voyage suggests, robots can go to distant or dangerous places that people cannot or will not readily visit. Robots are heading for outer space, the depths of the sea, the wastes of Antarctica, and the poisoned heart of radioactive accident zones. Some of the same robots are going to places to perform simple tasks that people just don't like, such as monitoring the same area for months on end. Whether exploring a burning building in a quest for survivors or watching a busy intersection for accidents, robots can survey areas faster, cheaper, and, in some cases, less obtrusively than human beings. Surveillance and monitoring are perhaps the most practical jobs for robots—some, like Sojourner, have already done them.

For these robots, autonomy is a key goal. For robots to function in faraway places, they must be able to act without supervision and be able to generate their own power. The Grail of remote-sensing robotics is autonomous action without a power cord. Unfortunately, generating adequate power over the long term is a huge technical problem—battery packs are not sufficient. In some sense, though, the technical obstacles may be a good thing. Precisely because autonomous robots function without a tether, they awaken people's fears of machines with the power to act on their own. The long buildup to these devices, and enlightening examples like Sojourner, may ease public fears.

Sojourner's enormous success created the opportunity for other rovers to make the move from earthbound research project to space traveler. Although further plans for Mars missions stalled after the two failed missions in 1999, research into robot rovers continues. We talk with rover team members Eric Baumgartner and Terry Huntsberger about FIDO and the Sample Return Rover, a smaller vehicle that will assist in bringing samples back to the landing vehicle.

Faith: Is your rover tethered?
Eric Baumgartner and Terry Huntsberger: No, nothing is tethered.
You said that in unison! If it's not tethered, what's that little thing that I saw behind it?
Huntsberger: Oh, the one in the lab here is tethered. You see a little white wire going back, but that is just the power cord and panic button.
Robot guys always jump on the word "tether" [laughter].
Baumgartner: Tether is a bad word. It implies remote control. The focus of our work is to create the technology needed for autonomous activities that last many days without human intervention.
Roboticists tell me the best-designed robots are simple, but how simple can a planetary rover be?
Baumgartner: It could be more complex. This is a simple design, actually. It is just one single CPU-board computer, and the rest of this is just what we use to interface everything—control the motors, read

Under the control of NASA engineers (above, from left) Eric Baumgartner, Hrand Aghazarian, and Terry Huntsberger, the Mars Rover robot slowly carries its small payload of rock debris and dirt up the ramp to its mother ship. The rover is scheduled to be sent to Mars on two missions, in 2003 and 2005. Once on the Red Planet, it will slowly roll across the surface for several kilometers, picking up a half-kilogram of pebbles, and then return to deposit its payload. The mother ship then will rocket itself into orbit around Mars, where it will be picked up by a French spaceship and returned to Earth in 2008. Because radio messages back and forth take more than eight minutes to transmit (when the Martian landing site is facing Earth), if the rover begins to fall into a canyon, it will hit bottom before its controllers know it. Since distance and planetary rotation make real-time control from Earth impossible, the robot takes images from stereo cameras to map the local terrain and picks the safest routes to way points.

the information back from the motors, take information from all the cameras that are on board. We've also made use of a lot of industrial control boards.

And these rovers are built to work on their own?

Huntsberger: The techniques that we use to control these vehicles give it what we call way points—locations to go in the environment. We can't talk to this rover in real-time—we have to send up a set of commands, wait almost a day, then get the data back. It has to do a lot of this stuff on its own. So we've developed stereo navigation techniques, generating local terrain maps and deciding whether or not a region is safe enough to drive in.

It also has a camera on a six-foot mast, which helps you and it survey the terrain.

Baumgartner: It's like you yourself are standing on Mars, looking around. There are two sets of stereo cameras on it—an outside stereo camera pair that's just black and white, with a wide field of view, for navigation purposes. This other pair has three filters in front of it, so you get a multispectral view. It's a much narrower view, so you get longer ranging, higher resolution imagery.

The rover itself is supposed to locate good sites for rock and dirt samples. How does it do that?

Baumgartner: We find interesting targets using what we call the remote-sensing suite, which is on our rover, then we'll utilize our *in situ* sensing suite,

which is on the instrument arm. There will be whole suites of instruments that will go out and investigate and sit up close and personal on rocks that you want to eventually sample. The mini-corer will position over the rock, pitch the core down, then drill into the rock and collect a sample.

That's a lot of things to do when you're so far from a repair shop. What happens when something fails? Do you have a lot in the way of backups?

Baumgartner: Very little. In flight, we are very concerned about mass, power, and volume. Power and mass are probably the biggies. The Mars vehicles will only have solar cells, so the power budget is fixed, based on the availability of solar instruments on Mars, which can degrade over time due to dust storms or dust collection on the solar cells. We really want to minimize power usage, so that drives us in terms of how many wheels can be driven at a single time, and things like that. A set of backups would be great to have, but we generally just don't have it because of those kinds of restraints.

What kinds of environmental problems are there for the rover on Mars?

Baumgartner: Radiation causes your computer to upset, maybe flip bits it's not supposed to flip, so some companies have built radiation-hard processors. Also, temperature range is a big issue. The temperature on Mars goes down to -100° Celsius at

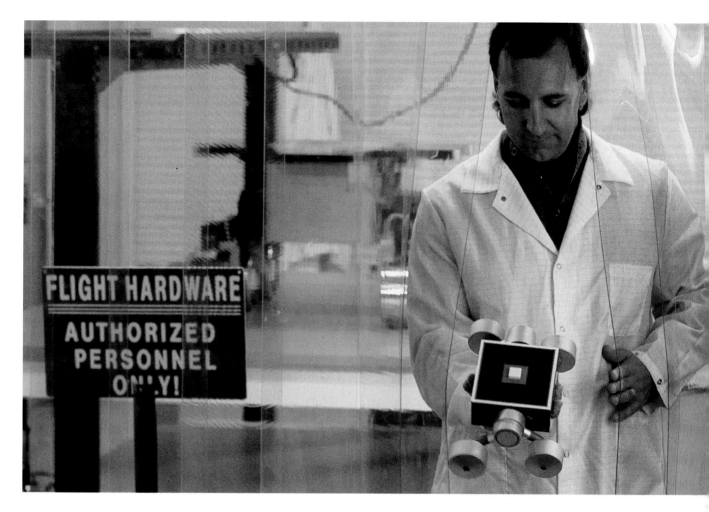

night, and so you sometimes have to operate at those temperatures until the morning comes and things typically warm up to about 0° Celsius.

The team here took FIDO for a field test in the Mojave desert—I read that it was quite a success.

Baumgartner: We had it out in the desert in October, and in turned out that it was 104° out there. We'd hope that the temperature would be a little lower than that. But we were driving along in a long traverse up this big ridge, and Terry [Huntsberger] was out there and he said, "The rover's on fire!" I almost thought he was joking. So they dropped everything and ran.

Huntsberger: Well, there was a puff of white smoke and we thought it was coming from the inside. [The solar panels caused the fire.] We had no fire extinguishers with us, nor are we used to this happening, so we ran and hit the kill switch. That stopped everything.

Baumgartner: It stops the motion of the rover. It doesn't stop the CPU, or the fire.

So it was the solar panels. Who put the fire out?

Huntsberger: Our lead guy, Bret Kennedy, did.

Baumgartner: He covered it over and put the fire out.

I guess you could have thrown sand on it.

Huntsberger: But there are a lot of optics.

So you try to minimize the sand throwing. I guess if it's going to happen, you want it to happen on Earth.

Baumgartner: Yes—but this probably wouldn't happen on Mars—this intense heat that we're dealing with. But the neat thing was that we hit the panic button, which shuts off the motion of the rover, but doesn't stop the CPU from running. It just disables motion. Generally, any robotic system is required to have a panic button so that you don't hurt anybody. So they pushed the button, put out the fire, and we basically restarted the motion control, and off it went again without missing a beat, though we had to detach the solar cells from the power system.

It doesn't go up with this other power system you had on there though—just the solar panels?

Baumgartner: Right.

This is the kind of project where there's a necessary marriage of science and engineering. How does that work?

Baumgartner: Well, we don't really know what a geologist does, and the geologist doesn't really know what robotics is, so it has been a marriage of the two. I've learned a great deal of geology from these guys—more than I'd ever known in the past, and they've learned a lot about the limitations and the advantages of robotic systems.

As an engineer, you must really like to see the actual application of your work.

Baumgartner: When it actually comes to fruition and does what you commanded it to do, or what the algorithm is doing—that's real special.

Working behind a plastic shroud that keeps dust out, NASA engineer Art Thompson of the Jet Propulsion Laboratory works with an early mock-up of what is called Nanorover, a lunchbox-sized space vehicle that will touch down on and explore a one-kilometer-wide asteroid. The small near-Earth asteroid 4660 Nereus is the target of a Japanese space mission that will launch in 2002. Carrying Nanorover, the Japanese spaceship will approach its target until it is close enough for the tiny robot to leave the ship and land on the asteroid. If all goes well, Nanorover will crawl over 4660 Nereus for about a month, taking pictures and collecting data. Because the asteroid's gravitational pull is about 100,000 times weaker than Earth's, the one kilo rover will move slowly and steadily, so as not to inadvertently launch itself into space. When its payload is full, it will return to the Japanese spaceship, which will in turn come back to Earth in 2006.

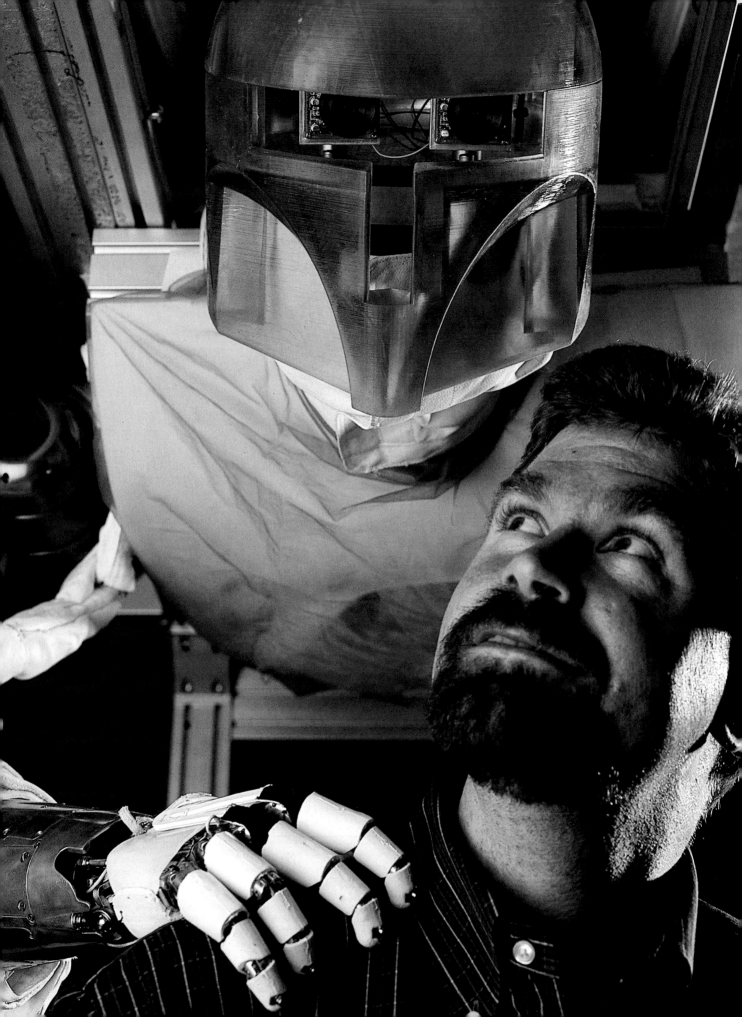

Space Cowboy

Two hours before you're supposed to pick up a malfunctioning military satellite, the satellite grabber in your payload bay is whacked by a piece of orbiting space junk. You desperately need to get out there and fix the grabber before the rendezvous, but preparing for a spacewalk is not a walk in the park. You have to put on your spacesuit first—and before you can do that, you have to sit for a couple of hours, breathing in enough pure oxygen that you can survive later in the low-pressure spacesuit. Unfortunately, you don't have a couple of hours—

Although this scenario may seem a bit far-fetched, it is precisely the kind of emergency that concerns the tele-robotics team at NASA's Johnson Space Center in Houston. Their solution? Provide a robot that can do necessary repairs, as directed by the astronaut, via remote control from the relative comfort of the spaceship. No muss, no fuss.

The research team is attempting to create a space robot that moves as deftly as a human being—a state-of-the-art humanoid that project leader Rob Ambrose says is unmatched anywhere else. Robonaut, as it is called, is elegantly designed, although the aesthetics are largely incidental. Its beautiful resin head, he says, is "just for show."

Faith: How would you characterize what you're building?
Rob Ambrose: What we feel we're doing is building the brain-stem level of control. The brain stem doesn't know anything about why it exists, what it is, or that it's part of the world. It just knows that it's being told to do something and it does it. And whether the command comes from a joystick, a person wearing a virtual-reality glove, or a bank of Cray supercomputers, the brain stem doesn't know what gave it that command. It just knows that it has been commanded to move and it's going to do it.
You're not just doing background research, but actually building a robot for space travel?
We have already tested two of the joints in the thermal vacuum chamber here at the Johnson Space Center and they performed extremely well. So this is the only space-rated humanoid on the planet right now.
Is anyone else trying to build something similar?
There are a number of space robots being designed but they're all very large. The next smallest one is just an arm. We went for the human package—the idea is to fit within the same EVA access corridors. [EVA stands for "extra-vehicular activity"—a spacewalk.] For example, Robonaut could ride sitting in a seat inside the shuttle. Robonaut could egress through the air lock with

an astronaut. It doesn't have to ride into space strapped down in the payload bay.
A lot of robotic manipulators have only a finger or two and an opposing thumb. Why does Robonaut have the human complement of four fingers and a thumb?
There's actually some logic to that. Robonaut can pull the trigger on a drill or pick up a flashlight while flicking the on-off switch with its thumb. What we were looking for was a mix of a dexterous set of three fingers above and a lower clamping set that can grab, so it can manipulate the thing that it's holding.
So it was just as easy to do five digits as four?
You have to have four. And the idea of five was to fit on the same tools that were designed for a human. Our theme here is to slip into the same form factor as humans and use all the same tools and equipment as an astronaut.
Is it harder or easier to build a robot for space than for Earth?
It's actually much harder to do it for space. And we didn't want to build just another lab robot. We're really looking to build that next step—a machine that could be spaceworthy and could get out there and actually do these tasks in low Earth orbit or beyond, maybe on the surface of Mars, or the Moon, or some other planetary body. Our motivation was to exploit the common denominator in all our existing space hardware—everything is designed for a human to service it. Everything is designed for a certain kind of capability to work with it.

If you were going to design something so a robot could take it apart, you'd probably make it much different. You would design it thinking about the robot that'd be working on it. We didn't want to do that, though, because we didn't want to redo the billions of dollars spent before fitting everything to human beings.
The whole process of getting ready to go outside for a space walk is pretty lengthy.
I've heard that three hours is the fastest you can get out because you have to prebreathe oxygen and get into that suit. The suit's a different pressure than the cabin. When you go into the lower pressure suit you have to be careful. That's the scuba diving part. When you're climbing in zero-G it's kind of like rock climbing but you don't have your weight.

The original space walkers had horrible experiences. They got outside and immediately started tumbling and wrapping themselves up in the umbilical [the tether holding the spacewalker to the spaceship]. If you look back to some of the first space walks, they were afraid that they were going to lose people outside. They got so worked up that they couldn't breathe right. They didn't think they'd get the guy back in.

Robo Specs

Name: Robonaut
Origin of name: Robotic astronaut
Purpose: Help humans work and explore in space.
Height: 1.9 m
Weight: 182 kg
Vision: Stereo cameras
Sensors: 150 per limb—position, velocity, torque, force, temperature
Frame composition: Aluminum (body padded with Kevlar and Teflon to make Robonaut fireproof and bulletproof—space junk is a big threat).
Project status: Ongoing
Information from: Robert Ambrose

Feeling a hand resting on his shoulder, Robert J. Ambrose (left) looks up to see a hovering Robonaut—the prototype for the robotic astronauts his team is building for NASA at the Johnson Space Center in Texas. Intended to accompany astronauts into space, Robonaut will be especially important in emergencies. Human astronauts need several hours to don their spacesuits and acclimatize their bodies to the suits' low air-pressure levels. As a result, they cannot quickly go into the vacuum of space to make an emergency repair. Robonaut, which needs neither spacesuit nor oxygen, should be able to leave the vehicle much more quickly than its human companions. Outside the spacecraft, it will perform its tasks under the control of a human operator at a tele-presence console. The machine's design is still being finalized—Ambrose installed the shiny acrylic helmet while his staff was still wrestling with the shape of the head.

The folks at the Johnson Space Center are proud of their machines, and rightly so. Their lab was packed with robots that worked pretty darn well, as experimental high-tech robots go. Because of the expense of house calls in space, Robonaut's mission description read like RoboDoc's. Robert Ambrose, one of the machine's creators, showed me a drawer full of medical instruments that it had practiced using, including a syringe for injections and laproscopic tools for minimally invasive surgery. The robot had passed its premed exams with flying colors—quite a feat for a second-generation robot bolted to the ceiling. Ambrose understood the importance of putting his machine's best face forward: he had designed an acrylic head in a style I would call Early Ambrose Droid. It is quite a leap, cosmetically speaking, from Robonaut's predecessor, DART. Bolted to the floor nearby, DART has a head that is a cross between a white baked-enamel appliance and a crudely drawn *South Park* cartoon character. DART was at ease when we visited, resting on its laurels. It was draped with a big brown sash declaring it a member of Chapter 152 of the South Texas Girl Scouts of America.

(From "The Mechanical Foundation of NASA's Robonaut," by Robert J. Ambrose, Christopher Lovchik, and Myron Diftler, for the Second International Symposium on Humanoid Robots, Tokyo, Oct. 1999)

The robonaut hand has a total of fourteen degrees of freedom. It consists of a forearm which houses the thermal vacuum rated motors and drive electronics, a two degree of freedom wrist (pitch-yaw), and a five finger, twelve degree of freedom hand. The hand itself is broken down into two sections: a dexterous work set ... and a grasping set.... Continuing up the arm ... a dense packaging of joints and avionics has been developed with the mechatronics philosophy. The endoskeletal design of the arm houses thermal vacuum rated motors, harmonic drives, fail safe brakes and 16 sensors in each joint. The roll-pitch-roll-pitch-roll-pitch-yaw kinematic tree is covered in a series of synthetic fabric layers, forming a skin that provides protection from contact and extreme thermal variation in the environment of space.

The breathing part won't be a problem for Robonaut, but what about its movements?

What the astronauts learned was—take it slow. Don't do wild motions because then you'll have to do another wild motion to stop and then pretty soon you're just hurtling out of control—take it slow. So Robonaut will very likely move in zero-G more like a tree sloth, with very slow, deliberate moves. That's the way astronauts climb, though you'll never hear them describe themselves as tree sloths, because that's not the persona you want.

Your colleague Fred Renmark [the primary operator of the Robonaut tele-presence equipment] is wearing a helmet that can see what Robonaut is seeing and a glove that can make Robonaut's arm and fingers move in concert with Fred's own. Describe for me what he's doing.

Fred's reaching out to grab a drill, which is tricky for a number of reasons. He can't really see the trigger on the drill, so he's going to try and place his hand so he can grip it with his pinky, ring, and middle finger around the butt end of the grip, with enough room for his index finger to be able to articulate the trigger. It looks like he was able to do that, even though he's essentially numb and can't really see.

Fred can't see very well because we've darkened the room for Peter's photography, but explain what you mean by numb.

He has absolutely no force-feedback. He has *some* visual feedback but because it's dark it's not even that good. As he was reaching out to grab the drill, he didn't really have any view of that trigger. He was able to place his finger in a really good spot because he has had more than 25 years' experience using his human hand and arm to reach out and grab things. Since Robonaut is so similar, all the years of instinct and training that Fred's had with his own arm came into play.

Now Robonaut, on the other hand, perceives sensations—exactly what its sensors are recording. It's able to sense its force, position, velocity, and temperature all up and down the arm but it has no way to communicate that back to its commander, Fred, sitting in the tele-presence garb.

Fred told me that in addition to his advanced degree from UC Berkeley, his years of intensive training in Donkey Kong *have really come in handy [laughter]. When you say "it perceives"—what do you mean by that?*

That its arm and hand—this upper extremity—has more sensors than any such limb that you've ever seen before. It has 150 sensor channels, which allows it to do things like cutting and stripping a wire.

And why aren't you using force-feedback?

Most of the advanced approaches to tele-operation aren't being tried right now in space because there's a lag in the technology from when you design some-thing to finally getting it flown. But Robonaut obviously is stepping right past that and going into a much more immersive form of tele-operation that we call tele-presence.

And so if Fred were sitting inside the space station or shuttle and operating Robonaut, would he be experiencing the same thing he's experiencing now?

Yes, but with the added experience of a free-fall environment, which is what you feel like when you're in orbit.

Really? My image of the zero-gravity experience is of a floating sensation.

It looks like they're floating but what they feel like is that they're falling. It's like that steepest spot on the roller coaster, when you're falling, but it's that way all the time. It's that steepest spot on the roller coaster for a week or two on the shuttle. That's why you get spacesick. Your inner ear tells you that you're falling but your eyes tell you that you're not.

A never-ending roller coaster ride. I have even more respect for astronauts than before.

Robonaut team member Myron Diftler was the lead researcher on the DART [Dextrous Anthropomorphic Robotic Testbed] robot, the early prototype of Robonaut.

Faith: You differentiate "tele-operation," which is operating something remotely, from what you call "tele-presence." Explain tele-presence.

Ron Diftler: Well, when someone wears the tele-presence gear what we try to convey to the person is that they become the robot. You're already used to using your own body parts; now we're giving you a new set of robotic body parts. We want you to try to take the same intuition you have with your own body, which is something subconscious for the most part, and use that with the robot.

I guess it was about two years ago we tried the mirror for the first time. We were trying to convey that information a little bit further and I brought the mirror up in front of DART's head and said [to the operator], "Here, take a look at your new face, now that you're using your new body." Most people have a momentary jump-back type response because they were all of a sudden seeing a face in the mirror that's not theirs.

Do people get used to it?

Many people who have come in to operate the system display some interesting characteristics. One gentleman from Australia was holding a tool with the robot [hand] and when he dropped it, he moved his own feet out of the way. Now the robot doesn't have feet and wasn't going to get hit by that tool, but basically you get so used to it after a relatively short period of time that your natural intuitions start to take over, even though it's a different body.

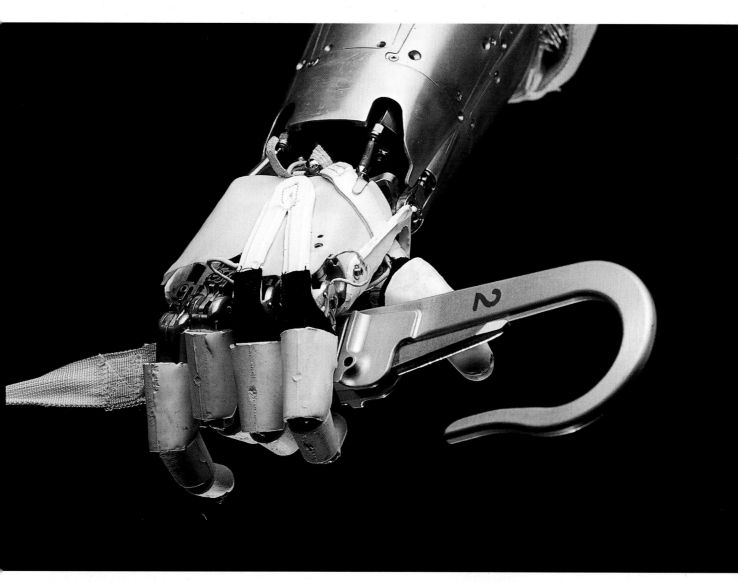

In a demonstration of mechanical dexterity, NASA's robot astronaut uses its hand (above) to open a tether hook of the sort that will be used during the upcoming construction of the International Space Station. Designed to be as humanlike as possible, Robonaut's hand has four fingers and an opposable thumb. Each joint has as many degrees of freedom as the corresponding joint in the human body—Robonaut can even tie a pair of shoelaces. It does not do this by itself, of course. Instead, it is tele-operated—it precisely follows the movements of an operator, who is guided by images from video cameras in Robonaut's head. In the first, developmental setup, the robot—then known as DART (Dexterous Anthropomorphic Robotic Testbed)—had its head (left) and hands joined to several long, immobile, pipelike cylinders. Later models (following pages) are much more closely patterned on the human body.

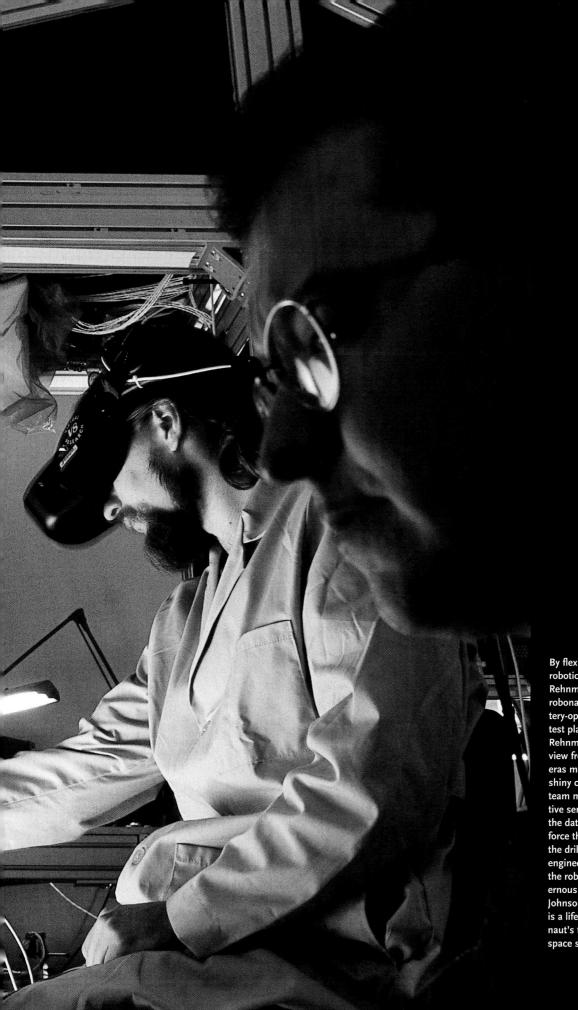

By flexing his data-gloved hand, robotics specialist Fredrik L. Rehnmark controls the NASA robonaut as it reaches for a battery-operated power drill on a test platform. Black goggles on Rehnmark's head give him the view from the twin digital cameras mounted in the robot's shiny carapace. In addition, the team may install pressure-sensitive sensors that will transmit to the data-glove a sense of the force the machine is exerting on the drill. Next to Rehnmark, engineer Hal A. Aldridge tracks the robot's test results. In a cavernous adjacent room in the Johnson Space Center laboratory is a life-size mock-up of the robonaut's future home: the NASA space shuttle.

Touchy Feeling

In 1990, the unified German nation created a new space agency, which described its most important goal as "increas[ing] scientific knowledge of the universe." Soon after, the German Aerospace Research Agency (known as DLR, after its German name), which conducts the agency's technical and scientific research, designed and built ROTEX, the first real space robot, which flew on the shuttle Columbia in April 1993. Remotely controlled by researchers on Earth, the ROTEX (Robotics Experiment) robotic arm captured free-floating objects in space. DLR is based in Oberpfaffenhofen, outside Munich, where we are spending the day with Max Fischer, postdoctoral researcher in computer science. Wearing a glove that transmits the motion of his hand to the robot arm's hand, Fischer shows us the latest version of the his agency's mechatronic systems arm and hand.

Faith: Are you talking specifically about a robot hand? You aren't also working on a full robot body?
Max Fischer: We are working on lightweight arms, which means the integration of hands and lightweight robots up to the shoulder.
What is the difference between the last version and this?
The hand you photographed [left] was called the DLR Hand. The new version will be called DLR Hand II. That's easy, isn't it? We learned a lot from the first hand. The main differences are a new actuation concept, better fingertip sensors and—the main point—the signal-processing concept. Sensor signals are digitized directly at the sensors and bus concepts are used up to the fingertip. This helped us to reduce the number of wires leading out of the hand from about 400 to fewer than ten.
What is the purpose of the robotic hand?

At the moment, the DLR Hand and the DLR Lightweight Arm are a test bed for research, but there are two mid-term goals we are aiming at. The first is to develop a lightweight robonaut arm for internal and external servicing in space. Such an arm could be teleoperated from the ground and could assist astronauts in performing tedious laboratory experiments or reducing dangerous extravehicular-activity time. The second goal is to provide a well-suited grasping and manipulation tool for service robotics. In this field, mobile platforms with arms and hands will do useful things one day. They will assist in fetching and distributing things in hospitals and maybe one day will run our households—robot hands could hold tools like a vacuum cleaner, or open doors, or operate a coffee machine.
You use commercial force-feedback gloves—that is, gloves that transmit back to the wearer some of the forces felt by the robot—rather than super-high-tech ones that you build yourself. Why?
Because they are available, robust, and sufficiently good. But there is a lot of work to be done in the force-feedback field to make things better.
Is force-feedback necessary for your manipulator to be effective? The NASA guys told us that good vision is more important.
Of course, vision is absolutely necessary for tele-operation. We use two pairs of stereo cameras when tele-operating our hand, one pair in the hand and one giving us an overview of the scene. We operated our hand without force-feedback when you and Peter visited our lab the first time. But having force-feedback for tele-operating a hand eases the task very much. You get a much more intuitive interface using force-feedback, because you feel the contact forces instead of having to estimate contact quality visually from the reaction of the grasped object.

Robo Specs

Name: DLR Hand

Purpose: Light, universal manipulator for space and service applications.

Length: 140 cm

Weight: Approximately 22 kg

Vision: Stereo vision

Sensors: Wrist force/torque, joint and motor position, joint torque

External power: 48 V DC, 20 kHz AC

KLOC: Hundreds.

Person-hours to develop software: Many, many...

Project status: Complete—next-generation hand soon operable, feels good, looks even better.

Information from: Max Fischer

Peter's Notes

We walked into the Bavarian village deli to buy a prop for a highly sophisticated robotic hand to hold. Thomas Borchert, the *Stern* magazine science writer, thought a huge white sausage would look good. I reminded him we were shooting a magazine story, not a porn video. To me, the big freshly baked pretzels with hunks of sea salt were picture perfect. Seven hours later we had overcome the technical difficulties of having Max Fischer and the robot hand both holding glowing light bulbs. When it was time for the pretzel shot I was shocked to find the bag nearly empty. Thomas and Faith had eaten all but one—we had worked through lunch. They rationalized it by saying they ate only the ugly ones.

Delicately handling a pretzel, the robotic hand (far left) developed at the Deutsches Zentrum für Luft und Raumfahrt (German Aerospace Center), in the countryside outside Munich, demonstrates the power of a control technique called force-feedback. To pick up an object, Max Fischer (left, in control room), one of the hand's developers, uses the data-glove to transmit the motion of his hand to the robot. If he moves a finger, the robot moves the corresponding finger. Early work on remote-controlled robots foundered when the machines unwittingly crushed the objects they were manipulating. Researchers realized that they were trying to operate robots that didn't have any sense of the force they were exerting—feedback of the type ordinarily given by the nerves in the fingers. Now that the robot is equipped with sensors, it can feed back signals to the data-glove—giving Fischer the sensation of touching the object, thus helping him handle it with appropriate delicacy.

Loco Motion

Robo Specs

Name: SARCOS Treadport

Purpose: To allow natural locomotion in virtual reality applications.

Length: 3.5 m x 2 m

Vision: Custom 3-wall CAVE display

Frame composition: Welded steel

Information from: John Hollerbach

Name: Dexter

Origin of name: DEXterous TElemanipulation Robot

Purpose: To place a dexterous robot hand on the end of a robot arm, expanding manipulation capabilities.

Creative inspiration: Hmmm, nothing too special.... I think I may have dreamed one night that we had a small, light hand on the end of the Adept arm and we were using it for tele-manipulation tasks. And now the dream has become a reality! In my dream I think Dexter was black— (Mark Cutkosky, principal investigator).

Finger length: 20 cm

Weight: 1.92 kg

Sensors: 2-axis strain-gauge force sensors on each fingertip

Frame composition: Aluminum

Batteries: None

External power: 115 V AC

KLOC: 25

Cost: $6,500 (components)

Project status: Ongoing

Information from: Weston Griffin

Seeming to touch the objects on his screen, Peter Berkelman (right), then a graduate student at the Carnegie Mellon Robotics Institute, scoops up virtual blocks with a special device that communicates the sensation of touching them. The device, which has a handle suspended in powerful magnetic fields, can move with all six possible degrees of freedom: up and down, side to side, back and forth, yaw, pitch, and roll. Used with special "haptic" software—the term comes from the Greek verb *haptesthai*, to touch—the device has force-feedback. That is, it lets users feel as if the handle were physically interacting with objects on the screen through a virtual scoop-shaped tool, sliding under the sphere or picking up the octahedron. Haptic interfaces like this one have become increasingly important as researchers seek to give people greater control in interacting with computer generated virtual or physically remote environments.

Weston Griffin plucks at the air with a hand covered by a special electronic glove. A few feet away, a robotic hand named Dexter makes the identical movement, picking up a small wooden block. Not only does Griffin, a Stanford University graduate student, control the robotic hand with his glove, he can literally feel the force when the robot hand grabs the block and adjust his grip accordingly. The glove is fitted with actuators that feedback to Griffin's fingers the sensations "experienced" by Dexter's fingers through its sensors as it manipulates objects. "When you get feedback like that," says Stanford's Mark Cutkosky, whose Dexterous Manipulation Lab we've invaded for the day, "you suddenly realize that the robot's fingers are your fingers out there, even though they may not look like your fingers." Cutkosky suggests I try it, and his graduate students agree to help. First they have to reset the computer controls for the glove.

Faith: Why can't I use the settings you are using?
Weston Griffin: It has thirty-one parameters to adjust.
Give me an example.
Ryan Findley: One is the size of each bone in your finger—there are seven bones we are interested in, and each one is different.
Michael Turner: Each person moves their hand in a different way, but the robot is always the same, so we need to adjust that mapping.
So, what are you doing to my hand?
Griffin: The first step is to put the glove on and move your fingers around together. The data is taken while you are doing that, and there's an algorithm that modi-

fies all of these parameters. Once we have a virtual model we can additionally modify the parameters.
Findley: There are two steps, calibrating your hand to the computer, and the computer to the robot. The first is quick and easy—the second isn't. It takes twenty minutes.
Turner: Each one of these twenty-two bumps is a flex sensor on the glove, measuring the bending of each one of your joints—everything from how your fingers spread to the curl of each knuckle. We are only interested in the sensors that are on the thumb and index finger because the robot has only two fingers. This exoskeleton-like cabling on the back of the glove is the CyberGrasp® device, which will apply forces to your fingertips. Put this on before we do the calibration because it will affect, somewhat, how the sensors read.
Findley: It affects how the glove fits your hand.
Turner: So this CyberGrasp® will apply forces to your fingers by pulling along on these cables. The motors are not strong enough to hurt your hands, and there are mechanical stops, so that they can't hyper-extend your fingers. You are perfectly safe wearing this device.
I'm now wearing the CyberGlove® and the CyberGrasp®.
Findley: That image on the computer screen is your index finger and your thumb. You can see when you put your fingers together in real space that they don't come together in the model on the screen—the glove fits your hand in a way different from the default fitting that we have set up in our model. Some of the calibrations will be closer to the default, and some of them will be wacky. Put your fingers together, and we will do that move-around process all over again. Now it's starting to chew the numbers a little bit—it's moving around and getting closer with each tick. Now it's grabbing a whole set of poses, where it records the angles of each of your joints.
It's incredibly complex. Who figured out the math?
Findley: It's a method developed by John Hollerbach [University of Utah] that we adapted. I'm going to tell you to grab some fruit: grab a cherry for me; grab an apricot. [Imaginary fruit to allow calibration of various distances between fingers and thumb.]
Griffin: Try and make these motions as natural as possible. Don't try to help us; let us modify the parameters. [*Weston tells me this in order to capture the different distance between my thumb and forefinger. Fifteen minutes later, the students finish the calibration.*]
Griffin: We're going to connect your motions to the robot's motions. Make easy movements until you get the idea—then we'll turn on the forces, and you'll see what those are. [Puts a block in the robot's hand.]
I can feel it tugging on my finger. Now I'm grabbing the block—I can feel it. Cool!
Findley: With some practice you can go from the dexterity of a two-year-old to the dexterity of a four-year-old, which is how I would characterize Weston, who is the most experienced at using the little robot [laughter].
It's bizarre, I feel disembodied.
Findley: It is kind of wild.

At the University of Utah, computer scientist John M. Hollerbach (above) puts a lab staffer on the SARCOS Treadport, a device that mimics the tug and pull of acceleration. Walking on a treadmill, the staffer is surrounded by a projected simulation of a Western mountainside. On a real hill, hikers must struggle with their own inertia to surmount the slope—a sensation no ordinary treadmill can provide. The Treadport uses force-feedback to push or pull at the user, uncannily evoking the sensation of climbing—a new dimension of realism for this type of simulation.

Force-feedback is widely used in data gloves, which send hand movements to grasping machines. The robot hand, which was built by the students (left) in Mark Cutkosky's Stanford lab, transmits the "feel" of the blocks between its pincers, giving operators a sense of how hard they are gripping.

Red's Rovers

Robo Specs

Name: Nomad

Origin of name: Desert wanderer

Purpose: To robotically explore regions like deserts, valleys, and moraines while mapping, searching, and making discoveries.

Creative inspiration: I sought a technological leap beyond the planetary exploration stereotype of robots as unthinking, tele-operated drones that were short range, short duration, and slaved to their lander for communication, computing, and human command.

Number of man-hours to build physical robot: 4 people, 1.5 years

Height: 2 m

Length: 2 m

Weight: 815 kg

Actuator type, number, and kind: drive and steer: 6 Brushless DC harmonic gearing motors

Vision: Panospheric camera for wrap-around viewing, high-resolution camera on a pan tilt for scientific and incidental viewing, 4 digital cameras in two stereo sets for wide-field navigation and automated safeguarding

Sensors: Laser scanner, high-resolution camera, panospheric camera, stereo cameras, spectrometer, magnetometer, compass, gyro, as well as sensors to measure tilt, chassis motion, steering and driving state, weather, impinging sunpower

Frame composition: Welded aluminum space frame

Batteries: 12 V lead-acid gel

External power: Fuel for onboard combustion generator

KLOC: Approximately 200

Cost: $200,000

Funding agency: NASA

Project status: Ongoing

Information from: William "Red" Whittaker

Peter's Notes

In January 1999, I had wanted to photograph Red with his famous robot Nomad, but at the time it was on a ship returning from South America after a 215-km trek in the Atacama Desert. We came back to CMU in July, and Red took us on a tour of postindustrial Pittsburgh— old factories, abandoned steel mills, slag heaps. We ended up by some coal piles down by the river in 100° heat—a test run for Nomad before it went to Antarctica to hunt for meteorites.

Red Whittaker's nova-bright reputation in robotics was the direct result of his success in building a trio of unmanned, tele-operated robotic work systems that in the 1980s helped to clean up the damaged nuclear power plant at Three Mile Island, near Harrisburg, Pa. The head of Carnegie Mellon's Field Robotics Center and of a robotics firm called RedZone, Whittaker has maintained a high profile ever since, building one rugged exploring robot after another. Half-ton robots Dante I and II descended into active volcanoes, the first in Antarctica and the second in Alaska. Whittaker sent a tracked robot named Pioneer to Chernobyl, where the broken reactor was far more radioactive than the one at Three Mile Island. It sits there now, waiting for the red tape to melt away. Perhaps his most startling creation was the colossal Ambler, a six-legged, four-meter-tall robot intended to stride over the surface of Mars (if, that is, NASA could have afforded to send such an enormous machine there).

When we visit the Field Robotics Center on a steamy weekend morning in July, we see Nomad, a Whittaker creation just back from a 215-km test traverse across Chile's Atacama Desert. In the heat of the Pittsburgh summer, Nomad is being winterized for another daunting mission: to conduct an autonomous search for meteorites in Antarctica's Elephant Moraine. Whittaker himself proves to be a tall, rugged, red-faced ex-Marine who looks like a farmer—and is. Having just flown in from Texas the night before, he is dog-tired and desperate for coffee.

Faith: You seem pleased with Nomad's exploits.
Red Whittaker: I don't pick favorites among my progeny. They're very different in their applications, very different in their purposes; they're very different in their technologies. As a body of creation, it's an unfinished work, but it fuels a revolution.
A revolution?
What I know is that a revolution has occurred. In too many ways, it is unrevealed, unrecognized, unseen.
Can something have occurred if it is largely unrealized?
There is a difference between largely unrealized and not realized at all. Did something occur when Einstein had the relativity insight? The exploration of the solar system is a growing enterprise. Excitement is afoot. A decade ago, the complete view of a space mission was a man in a can, a human-centric mission. It was the heroism and the romanticism of astronauts. And if you had an astronaut out there, then maybe Congress would back a mission for cold war interests. If not, then the activity just wasn't viewed as being viable. Fast-forward to 1999. Look at the missions that have just been declared—off to Mercury, bomb into a comet, explore the surface of

Europa [a moon of Jupiter] for life—and Pluto is on the horizon for a very distant destination. Nobody dreamed of putting a person in these missions. They're all robot missions.
Risky space situations are better for robots than for humans.
The transformation in large part is due to the visibility of the robot vehicle Sojourner exploring on Mars in 1997. Did people feel participative and involved? You had better believe it. Did it somehow inspire youth? Sure did, because kids were online logged in to the NASA website. In many diverse ways, it opened up a vision of robots as a workforce beyond the planet.

Now, you see, if we had started ten years ago promoting the image of enterprise beyond exploration, it would have been the O'Neill colonies—humans in bubbles, and that kind of thing. If you look at it now, everything from the art to the mindset is robots working in the interest and on behalf of the people. It doesn't matter if the enterprise is mining helium-3 for clean power for the Earth in the future, or beaming sun power down from orbit, or manufacturing materials and self-replication. We are creating a world with robots. Once generated, you can't take that back. One of my technical insights is that you just don't uninvent it.

For space missions, there are the issues of cost,

mass, and reliability. There are environmental implications that include everything from how you contaminate the atmosphere on your launch, to how you power in orbit, to how you reenter or become space debris. One of the things that goes along with crossing that line of manifestation in the real world is that if you haven't done everything, then you haven't done anything. And that's a tough game.

Give me an example.

I'm thinking now about Dante 1; that required conception, development, and deployment—I took a twelve-month gig. It was the notion of exploration of a live volcano for the collection of gases. Specifically, an inaccessible volcano, or one that

Not long before going to Antarctica, William L. "Red" Whittaker took a rare moment off from his busy schedule to accompany Nomad, his meteorite-hunting robot, on a practice run. The robot spent Antarctica's summer of 2000 on the ice, hunting for meteorites. Meteorites are preserved in the cold, dry air of Antarctica longer than anywhere else on Earth—more than 20,000 have been found there since 1969. With its onboard instruments, Nomad found and classified five—the first time that a machine autonomously made a scientific discovery.

Robo Specs

Name: Rosie

Origin of name: Robot character from the cartoon show *The Jetsons*

Purpose: Remote boom crane robot for radioactive environments

Creative inspiration: Three Mile Island and Chernobyl

Height: 2.1 m (boom stowed); 8.2 m (boom open)

Length: 2.6 m

Weight: 6,400 kg

Vision: 10 onboard cameras

Sensors: All motions closed-loop control with hydraulic power, resolvers for feedback, onboard status sensors (temperature, hydraulic fluid, pressure, etc.)

Frame composition: Aluminum, steel

External power: 480 V AC

KLOC: 10

Cost: Approximately $750,000

Project status: Complete (system deployed at Argonne and Oak Ridge National Laboratories)

Information from: Gary Dimmick

really called for some means of exploration other than human. And the coincidence is that it occurred a year after nine volcanologists got smoked in the 1993 Galeras Volcano eruption, in southern Colombia.

If you had a year in which a thousand firefighters died, robot firefighting would have a much better chance.

In the early going, it is no secret, I exploited that kind of circumstance, and I was a technical ambulance-chaser. A roof cave-in that killed or trapped a hundred people was a good occasion for a proposal. A Challenger falls out of the sky in front of millions of schoolkids. There are two parts of it. The first is that it is a robot game. Meaning, what can search for and recover all those pieces? I don't care if it is a 747 or a Challenger. Sure, there are people around, you see them on the boats, you can stick microphones in front of their faces. But if you are going to survey grids and do that sort of thing now, it is a robot game—it is transformed. But the other part is that those accidents are occasions for proposals, or for business deals, or for "grab your robots, and put them on a trailer," and you had better get on the road. To what extent does a robotics community, and I say that loosely, really act in a committed way, or understand that, or use that strategically? I shouldn't say "use it." What I am

after is the difference between talking about it and doing it.

Nomad went from the Chilean Atacama Desert to Antarctica. How were those missions related?

Now this loops us back to what we were talking about—motivations. ALH84001 is the name of an Antarctic meteorite from which someone intimated the evidence for the existence of life on Mars. [There were apparent traces of life in the meteorite, which was knocked from the Martian surface millions of years ago by a collision with a huge comet or an asteroid.] Very controversial, you know. It's like the nuke thing. I'm not pro-nuke or anti-nuke, I just do the robots. Me, I'm not big on "is it life, isn't it life?" But it's a reasonable context for getting to the Earth's poles, which are an analogy to the lunar poles, which are an ultimate intent. I cooked it up and tossed it in and got the nickel to build the machine.

You wait for an opportunity to build what's needed, and then you pounce on it. Persistence pays.

Persistence. Being able to hang in there. Too many of these guys fold before finishing. My grandfather used to sleep less than most robot developers do. I used to resent people who couldn't go for days, or had to eat.

Do you still believe that?

No. I don't correlate long hours with what occurs.
We were amazed to see how big Nomad was in real life. And how it moves.
It's important that you appreciated that and had that kind of reaction. It is hard to imagine that many people could view some robots for the first time, and not be enchanted, a little intrigued. And that's more predictable as robotic capabilities increase.
You want people to be dazzled by your machines.
There has to be a little surprise—in choosing toys or entertainment or betting on robots gaming against one another in the arena. It's important that people be interested. If people are not interested, there is no market.
What's the lifespan of Nomad?
Well, it's been in existence in some form for years now, going back to the desert trek. It is originally designed for a thousand-kilometer nominal life. When we outfitted it for the Atacama, then 10,000 kilometers. What I find is that many global robots find an enduring agenda as research or for some other purpose years and years and years after you have them. I don't get very attached to them in the long run.
I'll bet that Ben Shamah who actually did the hands-on building of Nomad does.

If he does, I'll have to break him of it. It's like farmers who care too much for their calves.
We were talking about robots in the home. What do you see happening in the home as far as people being affected by robotics?
One has to ask, "Do the economics of the home work for these robots?" I don't think so. "Does the culture of those homes work for robots' work?" In many cases, no. My cut on this one would be to go for the hotel trade. If you're going to clean room after room and make beds and scrub toilets, then let's clean and service hotels, not homes.
Now you're talking about repetitive tasks? Volume tasks?
In hotels I find myself looking at the extent to which all the doors are just alike and all the handles are a certain height off the ground. I find myself looking at the texture of the carpets, in part because they are fixed road maps with unique markings that are an exact map for every room in the place. To robots, all hotel rooms look alike.

Repetitive tasks were the simple tasks that made sense in the factory trade. When you're looking for winners in the early going, billions matter. Until homes can be standardized or robots become more competent, servicing homes makes little sense to me. This doesn't in any way diminish my sense of possibility or relevance for servicing homes. This could change the way we live, and that's what I'm all about.

When the Three Mile Island reactor (left, no steam rising from the abandoned cooling towers) failed catastrophically in 1979, the intense radioactivity in the plant prevented its owners from surveying and repairing the damage. Four years later, with conditions still unknown, Carnegie Mellon engineer William L. "Red" Whittaker designed several remote-controlled robots that were able to venture into the radioactive plant. The experience—and a similar, larger effort at Chernobyl—helped inspire Whittaker to develop other robots capable of acting in the field. In 1987, Whittaker and a partner started RedZone, a robotics company. Among its current products is Rosie (above), which can roll along concrete floors and stretch out to wield a jackhammer, drill core samples, or remove heavy equipment. Video cameras along the length of the arm transmit data back to a control center. In a practice session, an operator controls Rosie from a shrouded area that blocks his view of the machine, forcing him to interact with the robot via its cameras, thus simulating a real situation.

Radio-controlled outdoor mobile platforms, Micro ATRV (far left) and ATRV-2, are produced by Real World Interface, part of iRobot of Somerville, Mass. (ATRV stands for All-Terrain Robot Vehicle.) Their main purpose: to carry equipment in and out of areas difficult for human beings to navigate. Looking at the liquid-crystal display for the Micro ATRV, a Real World staffer directs it toward its larger cousin. Both of these highly agile robots can function in many, though not all environments. Although capable of slogging through deep mud with near-impunity, the ATRVs have to move with great caution on ice-crusted snow in this frozen glade near the company's production facility in rural New Hampshire —to avoid breaking through the surface and foundering in the deep powder beneath.

Eye Spy

"I'm trying to remember where we put it," Hagen Schempf says while rooting around in a cabinet for the grapefruit-sized rolling robot created by his research group at the Robotics Institute of Carnegie Mellon University and Shree Nayar's group at Columbia University. Shoving aside other robot parts, he finally unearths the robot, called Omniclops. "Almost like a baseball," he says. He's right; the robot is almost completely round.

"Let's see if I can turn this on." A quiet whirring noise comes from inside the sphere and it begins to roll. Schempf gives it a stop command but it keeps moving. "Where are the gains? Oh, there they are," he mutters. "It's hard to drive unless you're used to it." As the robot moves forward, we can see a blurry image of what it sees on a nearby monitor—just before the robot falls down a step. "Oops, I'm not the best driver," Schempf says, laughing. As he picks up this rolling bundle of vision sensors, I ask him what practical application he has in mind for Omniclops.

Hagen Schempf: What I really wanted to do was to turn this into a throwable baseball-sized sphere for urban conflict, stuffed full of sensors and cameras. Every American knows how to throw a baseball. It's not an exploding grenade—we wouldn't build one of those—but I'm sure the military labs will turn it into one. Yeah, they might short-circuit the batteries. Make it into a bomb! [Laughs]. I'm joking. But the idea was to gather data. I wanted it to be more than a baseball that just falls where it is thrown and sits there. The idea was to load a few on a larger mobile robot and pop them out.

Faith: You are very application-specific, aren't you?

I try to build systems not for exploring robotic technology per se but by applying it so that I can perform a useful task. CMU differs from most of the places that you've probably seen in that we try to do the extra step. We have more people trying to go one step further in terms of *real* applications but we don't hold the golden scepter. There are a lot of people catching up.

Do you think that robotics in general is maturing now as a serious field of research?

I wish we could make an impact with field robots—things that move around unrestricted—in the way that manipulator arms and stationary robotics have made [an impact] in car assembly and so forth. To me, I don't know where I would draw the line. Whether I would feel comfortable with a robot or not.

You mean living with one?

It's probably not going to happen in my lifetime. Mainly, even if the technology was there today I'd rather buy myself a Porsche than a robot that runs around and keeps my toilets clean, because, as brutal as it sounds, I can do it a cheaper way and get [a person] to do this.

That's what they were saying even in Japan, where the robot culture is so strong, about the high cost of robots versus cheaper human labor.

And I would say amen.

So building all-purpose machines is impractical.

Machines, just like humans, are optimized for a variety of things. You can't build a reliable and cost effective Swiss-army-knife robot. I don't believe that's feasible. Now Honda has built a humanoid [see page 34], which by inference is a Swiss army knife. But its behaviors are taught to it by humans. The shell looks human. If you squint, it's a very expensive human dressed in white going to a wedding. It is also too expensive and without the reliability the applications call for.

Robo Specs

Name: Omniclops

Origin of name: One-eyed monster, called Cyclops, in Homer's *Odyssey*

Purpose: To see what is occurring in areas hidden from view.

Creative inspiration: I wanted to be able to throw and roll a camera into areas just out of reach—hamsters in a pet store got me going.

Height: 15 cm diameter

Weight: 1.8 kg

Vision: Mini-camera

Sensors: Microphone

Frame composition: Aluminum, plexiglas

Batteries: NiCd

External power: None

KLOC: .5

Cost: $8,000 (including video transmitters and controllers)

Project status: Ongoing—using internal funds so development is slow and sporadic—developed Omniclops with Columbia University. It has a spherical camera and canted wheels to drive itself since then.

Information from: Hagen Schempf

Intended to provide 360-degree images of its surroundings, Omniclops, the robot "omnicamera," is being developed by Hagen Schempf (far left, holding Omniclops) of the Robotics Institute at Carnegie Mellon University. Schempf is now with the Robotics Engineering Consortium. Founded in 1994 with seed money from NASA, the consortium is located off the Carnegie Mellon campus and operates with great autonomy in this enormous facility. Behind Schempf on the main floor are autonomous forklifts; out of sight, other rooms are chockablock with robotic harvesters and mine diggers. The forklift, which can understand commands like "unload the truck in bay 4," should be deployed in Ford factories by the end of 2000.

A fifteen-centimeter-tall robot scout, Schempf's Mini-Dora (left) is intended to help police check out potentially dangerous situations. Unloaded from the back of a squad car, it could investigate buildings without risking the lives of police—as Schempf demonstrates by driving it up the front steps of an abandoned factory in a crumbling industrial section of Pittsburgh.

Sneakers and Seekers

Urbie the urban-reconnaissance robot is a scrappy 35-pounder with tracked wheels that iRobot, of Somerville, Mass., is developing as an urban field assistant for police and the military. The flat-bodied, stair-climbing, self-righting robot is designed to be a "smart" extension of a soldier's ability to operate in a changing environment. As portable as a toaster oven, Urbie operates both on its own and under remote supervision. Its cameras, sonar, microphones, infrared, and inclinometers make for an impressive sensing capability—and here we are throwing it over a strip mall fence.

Actually, we're tossing a well-worn demo model that does not have all the bells and whistles. Company software engineer Tom Frost told Peter that the robot is intended to be quite literally thrown into a situation. "We've done this plenty of times," Frost says now, as we stand by the fence.

After a few throws, wheel spokes begin to break off. Impressively, no one makes excuses. Instead, the engineers interestedly debate the cause of the problem amongst themselves, determining that a counterpart robot at the Jet Propulsion Laboratory has their proper wheels. "If they're not using them," Frost observes, "we should get them back." Later, Frost and Urbie's lead software engineer, Drew Bennett, talk about the robot.

Drew Bennett: The control of the robots, the artificial intelligence, is derived from behavior-based artificial intelligence. The idea is that if you look at the way that you or I would do things in day-to-day life, we tend to divide our activities up into little behaviors. You know how to brush your teeth. You don't really think about it because you have that all memorized. We do the same thing for the robots. We might have one behavior which tells the operator, if you are in a tunnel, just push the stick and go. The robot knows what you mean and it just runs down the tunnel, instead of having the operator focus her attention to make all the little course corrections.

Faith: In specific circumstances, you leave decision-making to the robot?

Bennett: Like a herder with a good sheepdog. He gives basic commands and the dog understands all of the details and does what is needed. The operator gives the commands "Explore the building, tell me about the first floor," and then does not worry about it until the robot comes back and sends a little message saying, "I'm done," and the map appears, and now you know everything you want to know about the first floor of the building. Something like that.

How would you define its level of intelligence?

Tom Frost: No smarter than I am, because I am programming it [laughs].

Bennett: I liken all robots to really quick two-year-olds—they do exactly what you say, not what you

mean. In Urbie's case, I'd say about its behaviors that it is probably as smart as an average dog. Overall, across all of its behaviors, it is probably dumber than any dog I have ever met. Because at least a dog has some clue as to what to do.

But once you give it the right set of information, it is reliable?

Frost: It is reliable for the majority of situations. There can always be something unanticipated.

Give me an example.

Bennett: Sonar is an example. Let's say that you go into a room where the walls are really smooth. It would be like you or I walking into a funhouse with one of those mirrored mazes. You keep walking into the walls, because you can't see them. The robot can't pick up the walls with its sonar, and it will just bump right into them, not realizing that a wall is there.

Frost: So in effect it is blindfolded, or worse.

Luckily there aren't that many funhouses around.

Bennett: We have no control over the environment at all—which makes it fun.

Shot-putting Urbie over a two-meter chain-link fence, Alan DiPietro (left), a staff researcher at iRobot of Somerville, Mass., shows how soldiers might use this remotely operated robot in urban warfare. Intended for surveillance, Urbie is a low-profile, remotely operated machine that crawls over obstacles on bulldozer-like tracks, beaming images of what it sees to its operators. In a simulated rapid-deployment mission from the comfort of a car, iRobot researcher Tom Frost (below) guides Urbie up a flight of steps. The robot is intended to be exceptionally durable, capable of flipping over and surviving shocks that would destroy most other robots. But the company still has a ways to go—one of Urbie's caterpillar tracks shattered when DiPietro threw it over the fence.

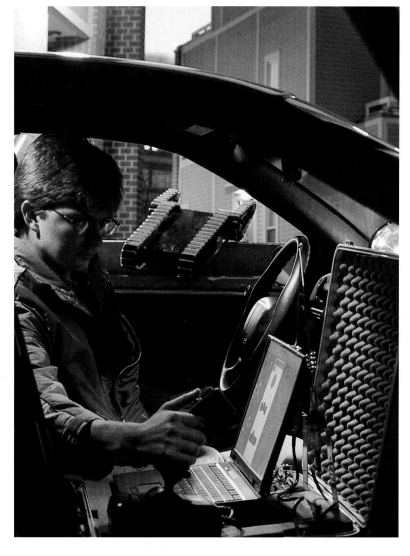

When a terrifying earthquake leveled part of Turkey in the fall of 1999, rescuers had trouble pulling victims from the rubble because it was too risky to crawl through the unstable ruins. As a result, some people died before they could be rescued. Shigeo Hirose of the Tokyo Technical Institute thinks he may have the solution: Blue Dragon (Souryu in Japanese). A light, triple-jointed robot with a digital camera in its nose, Blue Dragon could crawl through an earthquake-damaged building in search of survivors. Wriggling over a pile of shattered concrete on a construction site at the institute's campus, the battery-operated robot fell over several times—but righted itself quickly and continued slithering through the pile of stone. In addition to far-off places like Turkey, the machine may be useful close to home—Japan is both populous and earthquake-prone.

Sure Fired

By manipulating a joystick, I'm taking aim with a rifle that is located several feet away. This is not a video game or an arcade experience, but a remotely operated rifle system called a Telepresent Rapid Aiming Platform. TRAP T-2, as it is called, was designed and built by Graham Hawkes, a San Francisco-area engineer best known for his Deep Flight undersea-exploration-vehicle project.

TRAP T-2 is not a weapon itself; rather, it is a platform that allows an operator to remotely control a variety of devices, including rifles and sensors. Indeed, Hawkes expects that his main customers will be security and law enforcement officials who will primarily want the system for its potential uses in armed-conflict and surveillance situations.

In Hawkes' office and later in the field at a shooting range, the rifle seems spookily responsive—its aim remarkably precise. Hawkes believes, though, that his invention may actually reduce the likelihood that police will use their weapons. Because TRAP T-2 removes its operators from the direct line of fire, they don't have to respond viscerally to each potential threat; they have more time to think about when or if they need to pull the trigger. In addition, the system denies the opposing force the tactical use of covering

fire—the tele-operated rifle will not duck or run away.

Nonetheless, the operator's physical disassociation from the device, coupled with the joystick controller, the monitor that displays where the rifle is aiming, and the pinging feedback that gives the operator cues while tracking the target all make it feel like the video game it isn't.

Faith: How do you go from your Deep Flight project to tele-operated rifle?
Graham Hawkes: My work has always been in enabling humans to work in hazardous environments—deep sea, space, nuclear, etc. So, just as the deep sea can be a hazardous place because of pres-

sure, airspace punctuated by bullets is certainly a hazardous space. So, this is about remoting human capabilities into hazardous environments.
How does the rifle work?
The TRAP T-2 uses electric actuators to move a pan-and-tilt [frame] that operates the weapon mounted on it, and swings the weapon through 70 degrees. Two cameras—one wide overview and one view through the gun scope—are mounted on the system. The operator can therefore see a wide view or the view through the scope via a hand-held viewfinder or a full screen monitor. The operator uses a joystick to operate the system that uses a solenoid/electromechanical actuator to remotely pull the trigger of the weapon.
This is not a robotic rifle, as some have called it.
The TRAP T-2 is a remotely operated platform onto which you can mount small arms, sensors and surveillance systems. It is not robotic per se, as it is operated by a human being; a human still controls the actual firing of the gun, [rather than] a computer. The TRAP T-2 eliminates human error in firing the weapon, and so the weapon always shoots to its benched accuracy.
I know that SWAT teams have been practicing with it. What is the general consensus on its efficacy?
The Los Angeles Sheriff's Department and San Francisco Police Department both ran [tests], and San Francisco is still running long-term tests. General consensus is that the TRAP does exactly what [we say] it will do—eliminate human error in firing a weapon; enable stepped response from surveillance, to non-lethal, to lethal; keep police and soldiers and armed guards safe and out of the line of fire. We are focusing on military and government interest now and have sold several test units to a US government agency and to a US military agency.
Has a SWAT unit purchased your $47,000 package yet?
No law enforcement agencies have purchased TRAP yet. We are instead focusing on military and government interest. TRAP T-2 is the first stand-off system for military small-arms conflicts. For the first time, it brings decisive safety benefits down to the individual soldier level. Military and government agencies are interested in the TRAP T-2 for force protection, perimeter security of military installations, government/military facilities, border security, etc.
Do you have any foreign interest?
Significant—especially from countries which have contentious border areas.
Describe a scenario in which it might be used.
Protecting the perimeter of a royal residence or protecting a 200-mile demilitarized zone/border area. T-2s would be set up every 20 feet or so, and monitored from a central control room.
Is the heightened level of gun awareness affecting the interest in your device?
No. We do not make guns. We simply make the remote platform.

Robo Specs

Name: TRAP T-2

Origin of name: Telepresent Rapid Aiming Platform

Purpose: To save police officers' and soldiers' lives.

Creative inspiration: Graham developed a precise pan-and-tilt for an underwater filming platform. He then knew he could use this technology for a remote weapon system he had been designing in his head for many years.

Height: 46 cm

Length: 97 cm

Weight: 10.4 kg

Vision: 2 OEM cameras

Frame composition: Carbon fiber, aluminum

Batteries: 2 12 V

External power: 12 V

KLOC: 4

Cost: Undetermined, though Graham Hawkes sold his boat to pay for it.

Project status: Ongoing

Information from: Karen Hawkes

Designed for government agencies, military, police, SWAT units, and security forces, the TRAP T-2 (left) is a remotely operated rifle. In dangerous situations, lawmen could fire the tethered robotic gun from a position safely out of harm's way. Somewhat like playing a video game, the operator pushes a joystick button to fire the rifle. Using video cameras mounted on the side and top of the rifle, the shooter aims by moving the crosshairs onto the target. Developed by Graham Hawkes, a successful independent inventor, the gun is powered by a backpack worn by the shooter. At a shooting range near Hawkes's office in Port Richmond, Calif., the $47,000 rifle fires with deadly accuracy; untroubled by nerves or fatigues, it calmly and tirelessly tracks and hits objects no bigger than a quarter. A command and control unit with bigger video images and radio links sits on the tailgate of a pickup truck in the background. Some divisions of the US military are already using the system and several police departments are also testing the TRAP T-2.

Best Behavior

Robo Specs

Names: Ren, Stimpy, George, Miguel, Cartman, Sally, Shannon, Urbie, etc. ... can't remember them all.

Origin of names: Some are for explorers (Sally, Shannon). Some are for animation characters. Geroge the First is named after George P. Burdell, a fictitious Georgia Tech student.

Purpose: Basic research in autonomous systems. These were commercial stock robots that have been modified in house— we build software primarily, not hardware.

Creative inspiration: Psychological and ethological models of behavior

Project status: Ongoing, though some individual robots are finished

Information from: Ronald Arkin

Name: Silver Bullet and Bujold (following pages)

Origin of name: Beer (Silver Bullet was constructed at the Colorado School of Mines, which is right next to the Coors brewery) and a woman science fiction author (Lois McMasters Bujold)—all my other robots are named after women science fiction authors.

Purpose: Urban search and rescue

Creative inspiration: Marsupials like kangaroos and opossums

Height: 1.07 m (Silver Bullet); 10 cm down and 27 cm when erect (Bujold)

Length: 1.2 m (SB); 36 cm (Bujold)

Weight: 59 kg (SB); 2.3 kg (Bujold)

Vision: Dual digital cameras (SB); single digital camera (Bujold)

Sensors: Sonar, thermal, inclinometers, GPS (SB); camera, microphone (Bujold)

Frame composition: Plastic (Fisher Price Power Wheels) (SB); steel (Bujold)

Batteries: 8 12 V motorcycle batteries (both)

External power: AC wall plug

KLOC: 3

Cost: $38,000 ($20,000 for SB, $18,000 for Bujold)

Project status: Ongoing (second generation)

Funding agencies: National Science Foundation and DARPA

Information from: Robin Murphy

Unlike most of the other robotics labs we've seen, Ron Arkin's laboratory at the Georgia Institute of Technology is as neat as a nun's closet—the place even looks like it's been dusted. His dozen-or-so robots are tucked under counters, beneath desks, and in specially designed traveling cases. His students have given them cartoon names like Ren and Stimpy but any allusion to whimsy ends there. Arkin's great interest is building the inner robot. He leaves the general mechanics and the façade to others.

Faith: There are a lot of robots in your lab, but they're not the built-from-scratch sort that we see in other labs.
Ron Arkin: You'll see less, probably, in terms of new and glitzy robots here. You can call us stock-car racers, I guess. We buy stock cars [platform robots], modify them, then race them, so to speak. Our strength in this department is software development. We've got lots of robots—there is no shortage of robots here—but we work at developing intelligent behavioral control that transcends individual robotic incarnations. We develop what I call mission-specification systems, or software tools that are intended to enable average users—regular people—to use these systems without ever writing a line of code. We view that as important because the future of robotics lies in getting our work out of the laboratory and into the field, into the home—everywhere.

Do you consider yourself a roboticist?
Oh absolutely. I am a roboticist. I'm not sure exactly what that means, but that's what I am.

What do you think it means?
What it should mean is a person who is trying to make the field scientific, which is one of its weaknesses.

So you're saying it's not sufficiently scientific. How can it be made more so?
I was actually trained as a chemist a number of years ago and that brings a different perspective, I guess, than other folks'. If you look at computer science—and I am trained in computer science as well—there is often not as much science in computer science as one would like to see. Robotics to me is an experimental branch of computer science, mechanical engineering, and electrical engineering. It's a fusion of interdisciplinary studies, and it's very hard to master them all. So a roboticist needs to learn his limits and capabilities, in terms of the knowledge and expertise he has, and learn to rely on others to bring that additional expertise to solve the problems that need to be solved.

How did you go from chemistry to robotics?
It's not like I wanted to create an artificial human. It was just a nice natural satisfaction of my scientific curiosity. What I also liked about it was there was plenty of room to contribute. It was a wide open frontier. And being sort of a frontiersman I felt good

in that environment instead of in chemistry, where there were hundreds and hundreds of years of history in the field. You could make really important discoveries in chemistry but it's much harder. So the open frontiers of robotics excited me.

You have a funded project with Honda for their humanoid project [see page 43]. You don't have a Honda robot here that you play with, do you?
No, they tease me and say, "Do you want one?" I tell them no, because the technical support and maintenance would be prohibitively expensive. It would require a full-time technician. And you'd have to worry about them falling over. They have talked about in two or three years making them available to US universities. That's when they harden the design.

What are you doing for them?
We are working with the fundamental research group over there that is developing a hybrid behavioral architecture, not a subsumption architecture, which incorporated deliberation, motivation, and reactivity with the intention of eventually porting it onto the humanoid. We're initially fielding that in a Nomad 200 robot [built by Nomadic Technologies Corp.] and then using a tri-clops stereo-vision system and other things to reproduce Honda's laboratory facility here. The intent over the years, assuming we continue the relationship, is to get this architecture onto one of those guys [a Honda humanoid].

The Nomad robot is one of those round, trashcan-shaped robots?
Yes. It was actually discontinued, so we had to have Nomad build us one so we could have the same testbed as Honda's for preliminary development of the software architecture that would eventually be moved onto the humanoid. The Nomad serves as a preliminary waystation for a software architecture, integrating perception, action, deliberation, and motivation, while the humanoid itself continues in development.

Is the Honda robot any smarter than a real trashcan?
It's a phenomenal piece of equipment. Absolutely. But you're right, it's not very smart yet. And that's what we're working on—to take it to the next level. We're doing very interesting stuff with motivational control, with integrating representational knowledge.

You said it's important to get robots out to the masses. Are you working on anything consumer related?
Outside of Georgia Tech I'm doing consulting with Sony for the AIBO dog, although I can't talk at length about the sorts of things we're doing. We're hoping some patents will be generated shortly regarding those issues, but you know what I do and you know what they do and we're trying to do that in the context of robots for the masses.

The Japanese are really trying to become dominant in consumer robotics.
Just as we lost consumer electronics to the Japanese, we are becoming a third-rate nation industrially in

robotics as well. We have some great ideas. The world comes to pick up these ideas but unfortunately the bigger American companies have a relatively short vision. Of necessity—that's the nature of competition. Japan has MITI [the Ministry of Trade and Industry], which provides these large, long-scale, high-risk/high-payoff projects. The EC [European Community] can provide these long-range things. In the United States everything is much shorter fused, particularly from an industrial perspective. I haven't found companies here like Sony and Honda in Japan that are looking ten or twenty years down the road. The only people in this country right now who are interested in that and have the money to do it are the military. I lament the fact that industrially in many ways we will not be successful compared to what is going to happen in Japan and Europe. Except for military robots. But, we'll play in anyone's backyard but only doing open, unclassified research. That's where I draw the line, personally. Everything we do we can publish. I think it's utterly appropriate for a university to have that status.

Where do you see robotics ten years from now?
I think the field has sufficiently matured to the point that it will rapidly start reaching outside the laboratory and start to penetrate into the home and into industrial and private sectors throughout. I see the first wave of the invasion of robots as a nice benign invasion. This will be one where they will be in the service of people, all contingent on the cost of the manufacturer being ecologically competitive—they have to fit this competitive niche. It's not enough to build robots that do things—we can do that, but they will have to survive in competition with people or the other kinds of automation that are out there. That is the real question to be resolved, and that goes beyond artificial intelligence. I think they can handle that; it's a question of manufacture, packaging, reliability, durability, and usability. Those things will have the greatest impact on whether this stuff gets out of the laboratory. The issue is making these things work reliably, routinely, every day. I am not so confident that we will have this in ten years' time, but hopefully so.

Surrounded by the robots used in his Georgia Institute of Technology laboratory, computer scientist Ronald C. Arkin specializes in behavior-based robots—he's written a textbook with that name. Concerned more with software than hardware, he buys robots from companies and modifies their behavior, increasing their capacities. Industrial robots in factories work in controlled situations, he says, and so have little need to react to their circumstances. But outside such places, what Arkin calls "the physical situatedness" of the robot is "absolutely crucial" to its ability to act and react appropriately. Like many of his colleagues, he has been inspired by the way insects and other nonhuman life forms have adapted to their environment. But he also believes that most autonomous robots will need to think more intelligently than the organisms on which their design is based.

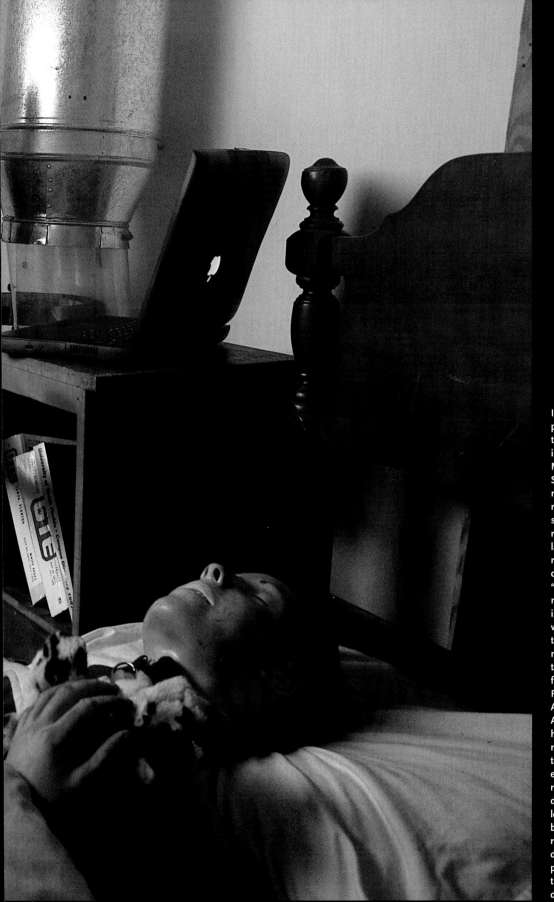

In a simulated bedroom complete with stuffed animals, tossed bedclothes, and a sleeping dummy victim, Robin R. Murphy of the University of South Florida keeps tabs on her marsupial robot—or, rather, robots. Developed to help search-and-rescue teams, the robots will work as a team. The larger "mother" is designed to roll into a disaster site. When it can go no farther, several "daughter" robots will emerge, marsupial fashion, from a cavity in its chest. The daughter robots will crawl on highly mobile tracks to look for survivors, feeding the mother robot images of what they see. Although the project is funded by the National Science Foundation and the Defense Advanced Research Projects Agency, Murphy's budget is hardly overwhelming—the mother robot is a Fisher Price toy souped up with off-the-shelf electronics components. Still, the researchers have learned a great deal. One of the biggest problems facing the project, Murphy believes, is the need for the robots' human operators to coordinate the simultaneous images provided from different directions as the robots fan out into a disaster area.

High Performance

The Gossamer Condor, designed and built by aeronautical engineer Paul MacCready in 1977, was the first human-powered machine capable of sustained controlled flight. Powered by a furiously pedaling pilot, the Condor, MacCready concedes, was very impractical, except for winning what was then the largest cash prize in aviation history—$100,000. "And," he says, "it got us and a lot of other people thinking more broadly." MacCready and his team followed the Condor with another human-powered aircraft, the Gossamer Albatross, which was pedaled across the English Channel and into the history books. If he lived in England, he'd probably have been granted a knighthood, but his own hard-won title is significant: MacCready is called "the father of human-powered flight."

But these accomplishments hardly exhaust the MacCready team's catalog of marvels, which includes his 1980 solar-powered aircraft, the Solar

A pioneer in aviation, Paul MacCready (above) designed the first human-powered airplane, the first piloted, solar-powered airplane, the first life-size flying replica of a giant pterodactyl, and a pioneering electric car. MacCready's firm, AeroVironment, is at work on the Black Widow (transparent model prototype in MacCready's hand), a remotely controlled plane capable of flying 40 mph for up to 20 minutes. Zipping along at treetop level, the 15-cm-long, 58-gram Black Widow could spot details missed by even the sharpest satellite cameras.

Challenger, the solar-powered Sunraycer, and QN Pterodactyl, a flying dinosaur built for the IMAX film *On the Wing.* "That got us very involved in servo-mechanisms, artificial muscles, and biological devices," MacCready says.

Though others with half his accomplishments might well be resting on their laurels, the 75-year-old MacCready continues to keep a workaholic's schedule. He talked to us about something called Technalegs™, and shared his thoughts about the future, and about autonomous robotic airplanes, some of them tiny, some that stay aloft in the stratosphere for long periods of time.

Faith: Could you have built a plane that stays in the stratosphere for long periods of time fifteen years ago, with the technology available then?

Paul MacCready: We tried to build one ten years ago, but the solar cells were not good enough, and all the power and electronics weren't efficient enough, so the project was shelved. It was reopened when it looked as though these technologies had become suitable.

We are also making little two-ounce airplanes that fly around with video cameras. Ten years ago we couldn't have made these. GPS [Global Positioning System devices, which use satellite transmissions to fix locations] ten years ago was too heavy. Now it just keeps getting smaller and smaller and cheaper and

better. And servos—we use motors that are a little bigger than a grain of rice but smaller than a pea. It's just incredible. Then you've got all the microprocessors that more readily do what you couldn't have done ten or fifteen years back, when, in the design process, all you had was a good systems engineer who just put logical things together; now you can employ elegant design techniques, such as multi-disciplinary design optimization with genetic algorithm.

These little planes are the ones you are building for the US Defense Advanced Research Projects Agency?
Yes. When you really care about tiny size the way the sponsor, DARPA in this case, does, the planes have to be smaller than six inches in any dimension. We end

up with things that have video cameras, telemetry, GPS, battery power for the propeller, gyros, servos—and the whole thing only weighs two ounces. That shows you what you can do.

It's quite a switch, going from human-powered flight to robotic airplanes.
I am lucky that I have been able to get rid of myself as president and CEO of AeroVironment [the company MacCready founded to pursue his projects]. I am chairman and founder, and do things that I am better at, while I have someone else handle the administration. That gives me a freedom that most people don't have.
With that kind of freedom, what projects are you working on?

Aerospace engineer Scott Newbern programs the flight computer in the Gryphon, one of the prototypes in the fleet of small robot jets under development at AeroVironment, a company founded in 1971 by inventor Paul MacCready. The Gryphon flies without a rudder or tail stabilizer. Instead, the operator controls the plane, which is powered by a miniature jet engine with many tiny, specially designed flaps that control the flow of air over its wings.

AeroVironment engineers (above, left to right) Marty Spadaro, Paul Trist Jr., Tom DeMarino, and Carlos Miralles cluster around the working prototype of the Mars glider, Otto. NASA sees an airplane as an important tool for exploring Mars early in the 21st century, and AeroVironment is seeking the honor of building the plane. Folded up like an origami figure, the plane will be packaged in a space capsule much like the one in Miralles's hands. When the capsule enters the Martian atmosphere, the plane will unfold itself and glide through the thin atmosphere, relaying data to an orbiting satellite that will retransmit the data to Earth. For use much closer to the Earth, AeroVironment is building the tiny Black Widow (right), which ultimately will be able to fly for an hour—or should be, if engineers can figure out how to pack more energy into its batteries.

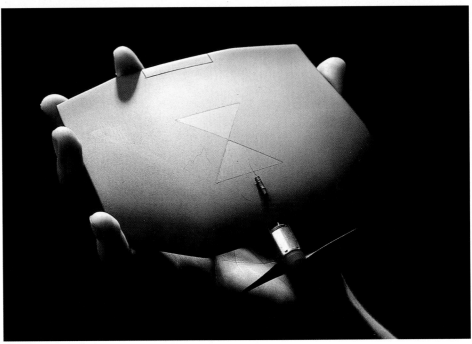

I am working on three devices to make exercise addictive. I do these on my own time, which I don't have much of, outside the company. If I had the company following all my whims, it would cease as a productive operation. Initially, I do quick developments with stuff that I get from a hardware store and some buddies that I pay to do some of the building. But then, if the device gets going properly, it gets sponsored somehow and becomes an official company activity. Only then can we turn it into a viable product. About 80 percent of what we do at AeroVironment is a derivative of these peculiar things that I have done on the outside, so I keep doing them.

What kinds of peculiar things?

Something that I'm working on right now I call Technalegs™. It is so simple—well, not all that simple—to try to execute in a practical device. If you look carefully at humans, you realize that we are superbly good in our versatility, but we are lousy when it comes to our musculature, skeletal structure, energy efficiency and so on.

With the Technalegs™, if you want to carry a big load in a backpack, let the backpack use its own legs but slave them to yours. Have its joints very close to your ankle and hip joints, and its leg lengths equal to yours. You provide all the stability and control. You can easily carry a load or a person on your back all day. You may get tired, but there is no problem in your control.

You are perfectly stable. You realize that you can forget the load in the backpack. You can transfer three-quarters of your own body weight into these exterior legs. Now, when you are walking, only forty pounds goes through your legs as a vertical load, so if you have lousy knee joints, or a hip that should be replaced in a while—suddenly it's as though you don't have those problems.

Over time, would your own body atrophy from not being used as much?

If you are helping yourself become a couch potato, you lose. But, if you help yourself become more active—"Hey, I couldn't hike or walk to work, now I can"—then you win.

Some people I've talked to, like Joe Engelberger [see page 186], expected robots to be a much more pervasive influence in our daily life than they are now.

The field of robotics at present is much smaller than I had assumed it would be. Years ago, imagining the future, I pictured robotic things having been around for decades. I though it would be a giant field. Of course, there is a fairly giant field of robotic devices for production lines, but if you go around to the universities and find good sharp groups working, you're able to catch up with what they're doing pretty quickly. Integrate all the research together and it's really not all that big at the moment.

Robotics is still a very young field.

And it's a field that will change the future of civilization in huge ways. I think that's not appreciated. It is this little bitty field—but realize that a lit match can be the start of a giant forest fire. The consequences can be big, but the beginning is tiny and you expect its growth to be easy to control.

You are quite concerned about managing this somehow.

In a business, you figure out what your goal is, clarify it, then work out and execute strategies that will lead you to the goal. It sounds logical, but the way humans actually operate is that they find what's interesting and useful to them now and keep pushing that—they don't worry about goals. Or they mistake strategies for goals.

It seems that technology engages people who might not be engaged otherwise, but I'm not sure this results in any long-term goal seeking. What goal do you think should be pursued?

I don't think that there is any one answer. But when I tried to decide what the most important phrase or question of the last one hundred years was, I had to say that it was probably, "I'm sorry, Dave, I can't do that" [laughter].

From 2001, when HAL the computer consciously refuses to release control of the mission?

Really, I cannot think of a more important phrase, and I am not the first person who has come to this conclusion. Seriously, it is inevitable that computers will have what we call consciousness as we move along.

That's inevitable?

I try to dramatize it enough so that people understand the challenge here. I have a certainty in my mind, which is a lousy forecaster because I am never right, that the surviving intelligent life form on earth will be silicon based, not carbon based. And that is why I think everybody should read *R.U.R.* [the first fictional look at robots, which portrayed them destroying humankind] and *War with the Newts*, which is the other comparable book by [Czech playwright Karel] Čapek.

So how do you fit into this scenario?

I label myself "an ambivalent Luddite at a technological feast." I think that expresses it. Here I am trying to figure out what people really are, and what they should be. What is desirable? What are humans? And at the same time working on technology, helping to advance technology, working with and utilizing technology.

I haven't really come to grips with it. And I'm not saying, "Hey, technology, go away." But I wonder whether we can keep it as a servant, of benefit to us, or whether it will become our master.

Robo Specs

Name: Gryphon (previous pages)

Origin of name: Mythological flying creature—half lion, half eagle

Purpose: Investigation of Modular Monolithic MEMS (M3) as an alternative to conventional aircraft controls.

Length: 1.4 m

Wing span: 1.8 m

Weight: 10.4 kg

Sensors: IMU, MEMS shear stress sensor arrays (~200 sensors total)

Frame composition: Composites

Project status: Uncertain—DARPA funding will end in a few months for first generation, second generation is in construction.

Funding agency: DARPA

Information from: Scott Newbern

Name: Otto (left, above)

Origin of name: Otto Lillienthal, glider researcher who inspired the Wright brothers

Person-hours to develop robot: The entire project, including design, fabrication, flight test and reporting, was accomplished over a 7-week period using roughly 1,000 man-hours.

Wing span: 1.6m

Weight: 4-7 kg

Sensors: 2 color CMOS cameras and RF downlink

Frame composition: Composite with aluminum folding mechanisms

Batteries: Ni-Cad

Project status: Successfully completed its objectives.

Information from: Carlos Miralles

Peter's Notes

I first met Paul MacCready in February 1987 in the middle of the Australian outback. He was with his solar car, which went on to win the transcontinental solar-car race I was covering for *Smithsonian* magazine. At sunset one day I asked him to snake a wire up his shirt so that he could cup a battery-powered pencil-light strobe in his hands. I knew he would like the idea of capturing light in his hands—he was already known as "the father of human-powered flight" and was the founder of a company whose motto is "Imagine the impossible—and do it!" In Japan, MacCready—a mild-mannered technical genius who only gets better with age—would be a living national treasure.

Auto Pilot

Robo Specs

Name: None

Purpose: Autonomous aerial support for various police and civilian tasks, including police chase, patrol, rescue, inspection, mapping, and filming movies

Height: 90 cm

Length: 4.3 m

Weight: 66.8 kg

Vision: Digital cameras (Sony) with DSP processor hardware (Texas Instruments), field-rate (60 Hz) image optical flow and region tracking (custom hardware), visual odometry (custom software and hardware)

Sensors: Dual Frequency carrier-phase differential GPS (NovAtel RT2), 3-axis rate sensor and 3-axis accelerometers (Litton LN-200), flux-gate compass (KVH), Laser rangefinder (Riegl)

Frame composition: Steel, aluminum, graphite, fiberglass

Batteries: 1 lead-acid 7 Ah 12 V

External power: 12 V (on ground only)

KLOC: Approximately 9

Cost: Approximately $250,000 for the handmade prototype; less than $60,000 if mass-produced.

Project status: Ongoing (possible deployment for oil exploration and forestry survey)

Information from: Omead Amidi

Deep in the basement of the Carnegie Mellon Robotics Institute, Omead Amidi adjusts the wing of the robot helicopter he is designing with Takeo Kanade, a Carnegie Mellon researcher who specializes in robotic vision. Several smaller versions of the project sit in his workshop. By integrating onboard cameras with gyroscopes and accelerometers, the scientists provide the helicopter's computer brain with the constant, precise measurements of position and attitude it needs for autonomous operation. To land and take off, the machine will also use laser rangefinders and acoustic sensors. Small and cheap to operate, the helicopter will trace a grid over the countryside, mapping what it sees below.

For centuries, human beings dreamed of flying through the air in mechanical devices. Omead Amidi's desire is to build a mechanical device smart enough to fly without a human being—an autonomous flying machine that can go from launchpad to destination without any supervision. In more technical terms, the Carnegie Mellon University researcher is building vision-guided autonomy into subcompact helicopters at CMU's Robotics Institute. After initially working on automobile controls, in the early 1990s he turned his attention to helicopters for his doctoral thesis , working along with institute director Takeo Kanade.

Faith: Why build self-piloting helicopters?
Omead Amidi: Basically, aerial imagery is useful for many things—high-speed chases, spying, filmmaking, inspection of power lines, rescue—many applications. It seemed worthwhile to build a machine that controls itself with imagery alone. And that's basically what's different about our machine. This can actually fly and produce visual feedback.

Even though it flies autonomously, can you take over control?
Absolutely. It has different modes. You can have some user interface. You can change what it's doing in mid-air. This particular helicopter here [points] has a human master—the human can put it in or out of computer mode. The computer can't decide to take over. Our new machine here [points] will decide. The computer will have the authorization, if something goes wrong with the human, to take over. That's the big step in that machine. It will actually take over and won't give control back. If you want it back, you have to type in a secret word and send it a signal. Only then will it give it back.

This is a retrofitted commercial Yamaha miniature helicopter. Show me how you've modified it.
This cup-shaped thing right under the "Y" in Yamaha, that's called an inertial measurement unit. It has instrumentation in there that measures angles and also acceleration. Directly below is an optical laser head. It shoots out lasers to this cylindrical thing that spins. What that does is scan the ground with lasers. Over here, you see a little box and that grabs the color of the object being reflected off the ground. As this thing flies, it makes a three-dimensional map of what's underneath. The mapping system is also used for positioning and for obstacle detection.

How might your system be used?
There is quite a bit of commercial interest in our machine—people in the electrical industry, the lumber industry, the fishing industry, the surveillance industry, and of course the military. The most important application right now is for high-speed chases. You get a drunk person driving at high speed and almost 100 percent of

the time there is an accident. Someone innocent is hurt or killed. The idea is to have these machines on rooftops in urban environments, just idling. They can start with a remote command and hover in the vicinity and give the cops images of the ground, saying, "Here's what we're looking at from above, or trying to chase." We have an object-tracking system. We can actually chase the runaway vehicle.

You're actually saying to it, "Chase that red car"?
The idea is that we will have a screen in a squad car, just a touch-screen, and we can point to what we want to chase. It's possible it will lose a target—for example, if it goes into a tunnel. It's not as intelligent as a human, who understands that it will come

out the other end. We'd like a human to be in the loop to say, "Oh, there it is again, chase that again." The idea is not to just have one, but five or six doing a job at the same time. You see, they're cheap compared to real helicopters!

How technologically savvy do the users have to be?
We would like to make it so easy that you don't have to know anything. In fact, I wouldn't be surprised if a five-year-old could easily maneuver this thing around, with our augmented control system. They always say the best interface is no interface. Basically, there is no interface. We just say, "Point to what you want."

How safe is it?
We don't feel comfortable flying this over people

yet, and the FAA doesn't have any rules about something like this yet.

Does that mean you can fly it?
No, we cannot, so we have to fly it on private property. We do follow the rules. We are looking actively for crash technology, deciding what we will do—explode some bolts to shoot out these blades before they land on people, or maybe something else. We don't care about the craft itself, just what's on the ground or what's around it.

What if you're flying near power lines?
It has a very sensitive laser system that will find wires. So those things are not an issue. What is an issue is if our machine is out of control and hits other aircraft or falls on top of somebody.

Work mates

In a photo-illustration depicting one way that roboticists are developing robots to take over routine tasks, the Electrolux Robotic Vacuum Cleaner sucks up hair at a Hamburg barbershop. Soon to be available from Electrolux, a Swedish appliance company, the machine will vacuum floors constantly with its small, quiet, battery-powered motor.

Machine Maid

Robots found their first real-world application on the factory floor, and heavy industry is still the environment in which robotics plays its most important role. The industrial robot dates back to the late 1950s and is largely associated with the entrepreneur and innovator Joseph F. Engelberger (see page 186), whose work with inventor George Devol foretold the future of mechanized industry.

It wasn't until the Second World War, when military researchers developed servomechanisms to aim weapons, that the technology to create precisely mov-

ing robots became available (see servo, page 235). Soon after the war, Engelberger realized that the techniques could be applied to robots. He and Devol, who had applied for patents on servos, designed and manufactured Unimate, the first industrial robot.

Simple by today's standards, Unimate was first installed in a General Motors plant in 1961. It had a robotic arm that picked up parts and placed them on a conveyor belt. Unimate could not respond to its environment—it was capable only of precisely repeating the behavior that had been programmed into it. To create those actions, Engelberger and Devol placed the robot on the factory floor and

moved it through its assigned tasks, recording its exact physical location precisely. The robot then played back its actions like a tape recorder.

These robots found their first widespread use in the auto industry, which had an ideal combination (for robot builders at least) of high labor costs, repetitive tasks, and conditions that were dangerous to human beings. The machines soon spread to food-processing plants, electronics-manufacturing centers, and almost any factory that did a lot of repetitive welding. The advantages were obvious—robots are much more accurate than people and never get sick (though they've been known to spring a hydraulic

leak or two). Indeed, manufacturers began building what are called "dark factories"—robotic assembly lines without lights, heat, air-conditioning, or toilets—staffed entirely by untiring machines.

The initial success of industrial robots was so overwhelming that in the 1960s many roboticists thought their machines would quickly move from the factory floor into people's homes. In fact, a trial-version robot lawn mower or two did, but the dream of a robot butler in every dining room proved to be just that. It was too difficult to develop robots capable of negotiating the complex environment of a home.

Even in factories, robots did not always have a clear advantage. In competitive businesses, it was often smarter to build a factory in a country with cheap labor than to invest in costly machines. As the fortunes of the auto industry—long the biggest market for robots—rose and fell, the fortunes of robotics companies rose and fell with it.

Today, industrial robotics is still not a huge business. Total annual sales, worldwide, hover at about $4.2 billion, considerably less than half the annual sales of, say, compact discs. Nonetheless, working robots are gradually spreading, gradually improving, and gradually moving into new areas. They might even invade the family homestead and stay this time.

Robotic pool-cleaners have been around for some time and companies are field-testing roving personal assistants, robotic vacuum cleaners, and lawn mowers. More important, robots are also beginning to appear in hospitals (see page 186). With practice, surgeons, using robotic hands, can now operate more precisely than with the human hand, which always shakes ever so slightly (see page 175). And doctors are using robotic vision systems to establish precise configuration when they implant artificial hips in patients (see page 177).

Even as these systems come into wider use, the prospect of robots becoming part of our daily life is unsettling for some. One source of this instinctive dread is easy to understand. Because the popular image of robots is of humanlike creatures, the notion of making them is, to some, tantamount to an act of hubris. To make a human being, even an artificial one, is to assume the role of the supreme deity, an action that in the Judeo-Christian tradition inevitably precedes a fall. In the first work of literature featuring robots—*R.U.R.*, a 1920s-era play by Karel Čapek—the soulless machines annihilate the human race.

In Japan, such fears play a much smaller role. Whereas children in Europe and the Americas have long grown up with images of Frankenstein's monster, Terminator, and the golem, Japanese children have been raised with Mighty Atom, a benevolent cartoon robot (see page 197), and other cartoon heroes. This is one reason that the robotics industry is booming in Japan—and a reason that the idea of a future filled with helpful humanoid robots enjoys such popularity.

Robo Specs

Name: Kärcher 2000

Origin of name: Company name

Purpose: Facilitates cleaning duties in private households.

Height: 9.3 cm

Length: 280 mm

Weight: 1.4 kg

Sensors: Robot is still a prototype, therefore I can't give you at the moment all the technical data.

Person-hours to develop project: 8 1/2 man-years for hardware, 3 1/2 man-years for software

Project status: Market launch planned for January 2001.

Information from: Frank Schad

Peter's Notes

The photo on this page was shot for the German magazine *Stern*. I had already shot the Electrolux RoboVac in Hamburg but the editors thought the Kärcher vacuum looked better, and it was German-made. So we flew to London, where we were met by a Kärcher executive who had flown in from Munich hand-carrying his company's RoboVac prototype, and a photo stylist, who had rented a house, a dog, a rug, and a middle-aged model. Five hours of lighting and styling later, the model was smoking a cigar in front of the fireplace, watching the telly, with the family dog in his lap. An assistant strategically threw dirt and ash on the rug at a point just below the cigar, causing the RoboVac to go back and forth over the spot until it was clean. No, the dog's name was not Spot.

Sucking up ashes in a London living room, the RoboVac—shown here in a photo-illustration—shuttles randomly around the area, vacuuming everything in its path. Built by Kärcher, a German appliance company, the RoboVac monitors the level of dirt in the stream of incoming air with its optical sensors—that is, it detects when an area especially needs cleaning. When the RoboVac hits a grimy spot, the machine passes back and forth over it until the incoming air is clean—and so too, presumably, is the floor.

Home Bodies

Robo Specs

Name: Personal Robot R100

Origin of name: R for "robot,"—100 does not have special meaning but I can say that it comes from NEC's 100-year anniversary.

Purpose: To verify the possibility of personal robots.

Height: 44 cm

Weight: 7.9 kg

Vision: 2 digital cameras

Sensors: 3 microphones, touch sensor, ultra-sonic sensors and environmental sensors

Frame composition: Plastic

Batteries (type, duration): 24 V rechargeable, 1.5-2 hrs

External power: None

Phrases recognized: About 100

Project status: Model completed, next generation in planning.

Information from: Yoshihiro Fujita

Peter's Notes

I was prepared for some pretty stupid robot pet tricks when we arrived at the NEC research headquarters outside Tokyo. Their public relations person had couriered a tape to our hotel and we had watched a very cheesy video of their little *South Park*-like characters in action in a cheap living room mock-up. At NEC, we were met by the leader of the project, Dr. Fujita, who was very young (by Japanese standards) to be heading up such an undertaking. We were pleasantly surprised by how quickly and cutely these smart little creatures did the mundane tasks they were voice-commanded to do: turn the lights and TV on and off, control the channel and volume, do a charming little song and dance. The uncharming part was the demo room. It was the same cheesy living room mock-up we'd seen in the corporate video, partitioned off from a much larger room by a beige curtain. Looking to improve the background for a photo, I peeked behind the curtain—there was a huge room full of engineers at workbenches stacked with equipment. A much more interesting room for a photograph. Jokingly, I asked if we were in Oz with the Wizard behind the curtain controlling the robots. "No," they answered most seriously. I asked if they would pull the curtain back part way to show the research. "No," they answered most emphatically, "Company secrets, new version."

Executives at the staid electronics firm NEC were reportedly shocked by the onslaught of media attention that a brightly colored research project attracted, even though it was under development in a back room of the company laboratories. A burbling, toddler-sized personal home robot on wheels, R100 is the brainchild of a team of young Tokyo-based NEC researchers who have been working on it since early in 1997. The machine is intended to be an unthreatening interface for many company-developed technologies, including voice recognition, visual recognition, and mechatronics (a combination of mechanical and electronic engineering). "We had all these interface technologies," says company spokesman Aston Bridgman. "Why not put it into the ultimate interface?"

On our visit with R100 in a simulated living room at the laboratory, project manager Yoshihiro Fujita shows us how it turns a television on and off and switches to specific channels. The machine can recognize people by sight or by their voices, can deliver messages from one person to another, and can serve as an Internet connection, Fujita tells us. It can also dance, something of a specialty in Japanese robots. Long after we leave, the frenzied little ditty it sings while dancing plays over and over in my head.

What must still be done before R100 goes on sale? One problem, Bridgman says, is that TV remotes use different standards. That's changing, he says, "at least in Japan." Japanese companies are rushing to wire together every aspect of the house. This will let robots like R100 become a sort of rolling remote control for everything electrical and electronic in the house. People in the living room will be able to summon the robot and ask it to turn down the oven in the kitchen, switch off the porch lights, and tell Junior to keep working on his homework. As Peter sets up for the photograph, we listen to the conversation between a researcher and the robot, which speaks in Japanese in a techno-Mickey Mouse falsetto. Our interpreter translates:

Technician: Good morning. R100 robot: Good morning! *Put the television on.* [It doesn't.] Do you want me to do something else? *Put the television on.* Okay! [Turns on TV.] *Put the light on.* Okay! [Turns on light.] *Put it on Fuji [TV network]. Put the volume up.* Okay! [Changes channels.] *Put the light off.* Okay! [Turns off light.] *Put on the weather report.* Okay! [Changes channels.] *Turn the television off.* Okay! Anything else? *Will you go to Grandmother's house on the weekend with me?* The power is off [no answer about Grandma]. *Dance.* Okay! [plays music, spinning around in circles] *Bye-bye.* Bye![Rather indignantly.]

Faith: You're working under the premise that a robot like this could be a desireable purchase for consumer? Yoshihiro Fujita: We want to find out if this kind of robot will be acceptable. Until recently, this kind of robot was only in science-fiction movies. But as the technology becomes more advanced, it is time for this kind of robot to become practical in the home. So we are not focusing on the mechanism as much as we are focusing on the human interface. *When you say you're not focusing on the mechanism, you mean in terms of more sophisticated mechanics?* It is not time yet to make something like a walking humanoid, but if it is very simple, like the R100, and safe enough, maybe we can start using this.

That is what I want to find out.

The lights around the eyes—do they turn on for a reason?
There are two reasons. One is for expression. The other is that they indicate to the person whether he is detecting you or not.

The robot recognizes you—how does it do that?
It uses the images it has previously recorded and stored.

Do you have to be directly in front of it for the robot to record your face?
Yes—then, if you call him, he will find you and approach you. He will call your name and ask you if you want help with something, or give you something like a message. When he has nothing to do, he will wander around the room.

[Meanwhile, Peter tries his hand at talking to the small robot, though it only speaks Japanese. As he speaks, it zips around on the floor in front of us, and stops in front of him.]
Peter: Hello. R100 robot: Hi! *How are you today?* It turns on a light, but says nothing to him, as it doesn't understand English. Then it says, "Bye, bye" and motors around the room, singing to itself.

Though its vocabulary is limited, its charm decidedly is not. When voice recognition software matures and video recognition is perfected, and if its voice comes down an octave or so, R100 might not be too annoying to have around.

Robot designer Yoshihiro Fujita stares into the electronic eyes of R100, his personal-assistant robot. The robot can recognize faces, identify a few hundred words of Japanese, and obey simple commands, but its most important job, Fujita says, is to help families keep in touch. If Mom at work wants to remind Junior at home to study, she can E-mail the robot, which will deliver the message verbally. To take the sting out of the command, the robot can sing and dance—a charming feature that is one reason NEC is inching toward commercializing the project.

Street PRoPosition

Robo Specs

Name: PRoP

Origin of name: Short for Personal Roving Presence.... Since my interest is in using these devices for social communication and interaction, assigning them names is problematic. Naming a PRoP should feel as strange as naming your telephone. Names provide an unwanted labeling and are almost always associated with some gender and/or nationality. ProPs should always project the identity of the remote user and hence maintain a neutral presence themselves.

Purpose: Tele-presence for the masses.

Height: 1.5 m

Length: 45 cm

Weight: 20 kg

Vision: Canon VC-C3 pan/tilt/zoom camera

Sensors: Odometry encoders, 7 ultrasonic sonar range transducers

Frame composition: Aluminum base, ABS plastic body

Batteries: 12 V 10 Ah lead-acid

KLOC: 12 (approximately)

Cost: $10,000 (approximately)

Project status: Ongoing—this research is an extension of work originating back in 1995 on space-browsing helium-filled flying blimp Internet telerobots with cameras and microphones

Information from: Eric Paulos

Berkeley graduate student Eric Paulos (right) calibrates his Personal Roving Presence (PRoP), which he describes as "a simple, inexpensive, Internet-controlled, untethered telerobot that strives to provide the sensation of tele-embodiment in a remote real space." In other words, Paulos is trying to build a kind of avatar—people could dispatch it to distant places to represent themselves in, say, business meetings. Using audio-video links, the operator could move the robot about and converse with people there. To reinforce a point, the robot can even make simple gestures with its pointer. On a test run (far right) on Telegraph Avenue, a busy retail street near the UC Berkeley campus, Paulos controls the PRoP from a short distance away, via a remote link. Most passersby react to the gangling machine as it were a friendly novelty item from the next century. One amused man in a wheelchair even stops and asks it for a light.

Every day, many strange things roll and stroll down Telegraph Avenue in Berkeley near the University of California campus. While we are there one mild winter afternoon, a small battery-powered robot sporting a small flat video display screen and a pan-tilt camera stops to chat with local street vendors, fortune-tellers, incense-sellers, shoppers, and students. The conversations are actually with the machine's designer, Berkeley PhD student Eric Paulos, who, though he is designing the system to be operable over the Internet, operates it today from a remote-control station a short distance away. Whether it is rolling into the local body-piercing shop or the Mrs. Field's cookie store, people react in the same way—staring, laughing, shaking their heads in amazement. Paulos couldn't be more pleased. Far from being strange, he explains, the robot—he calls it PRoP (Personal Roving Presence)—is doing precisely what it is supposed to: acting as a kind of ambassador. I ask him what he means.

Eric Paulos: It's personal tele-presence—a simple sort of device that allows people to project themselves into a remote space. Unlike NASA, where you go and suit up and have special gear or heavy, expensive hardware, this is tele-presence for the masses—anyone with an Internet-connected computer, standard AV hardware, and a joystick is ready to go.

Faith: Regular videoconferencing is where you talk with someone through a video link. How is this different?

The PRoP emphasizes the importance of real physical presence in human interaction. It allows a remote user to have a physical body, which makes a huge difference because you can actually move up to people and interact with them. You are not stuck on a desktop.

Because PRoP has a microphone, you're able to hear what's going on and respond.

I should be able to, if it's not too windy. The PRoP has at least two-way audio and video. Hopefully, with the live video of my face projected onto its screen, people won't mistake it as some kind of strange android police surveillance robot. The PRoP augments current remote communication techniques. Obviously, it's not designed to replace real human interaction.

You purposefully didn't anthropomorphize PRoP—no right and left cameras in the head positioned to look like eyes, no arms or hands.

Right. The problem is, that if it looks too much like a human, the whole thing takes on a creepy feeling and that shatters the goal of facilitating human interaction. It does have some humanlike elements. A simple pointing device placed sort of near where one might expect a hand is great for pointing things out and gesturing to people. Perhaps a laser pointer might help.

A laser pointer might frighten some people.

Yes, safe cohabitation with humans is most important. The body is mostly built of plastic. The idea was not to use a lot of metal, which would make it look robotic, because essentially it had to be something that was nonthreatening to people. It almost had to look flimsy.

You've succeeded.

The idea is that, just the way you use the telephone, you pick up a phone, you say, "Oh, I'm talking to Joe." You're not really talking to the phone. It's just this piece of media in between. So this is kind of the same thing. However, we have a long way to go. We will need acceptance of PRoP as just another piece of media between remote people just as the telephone is accepted today.

With government-funded construction clogging roadways throughout Japan, traffic robots like this one have become increasingly common. Standing in the Tokyo restaurant-supply district—chockablock with the artificial sushi exhibited in Japanese restaurant windows—this artificial policeman politely raises and lowers its arm to slow down approaching vehicles. Most traffic robots are simple cutout figures with battery-powered arms, but some—one popular variety is called Anzen Taro (Safety Sam)—are full 3-D mockups of policemen so realistic that oncoming drivers can't tell them from the real thing. Although the makers of these machines describe their products as "robots," many engineers would not, because they do not respond to their environment and cannot be reprogrammed. But the manufacturers may have a point—in performing its single, repetitive task indistinguishably from a human traffic policeman, Safety Sam and its ilk clearly pass the Turing Test (see page 236).

Test Byte

Robo Specs

Name: Waseda-Okino Jaw No. 1

Origin of name: Joint work of Atsuo Takanishi Lab at Waseda University and Okino Industries, Ltd.

Purpose: To clarify the human mastication system from the robotics point of view and to develop a basic mechanical and control structure for mastication science as research and education tools.

Creative inspiration: No other mastication robots in existence.

Height: 51 cm

Weight: 10 kg

Sensors: 9 linear potentiometers; 9 tension sensors; 10 strain gauges (temporomandibular joint force); micro pressure (biting force)

Frame composition: Duralumin

External power: 100 volts AC

KLOC: 8.3

Cost: $100,000

Project status: Ongoing

Information from: Atsuo Takanishi and Hideaki Takanobu

Peter's Notes

In the robotics lab at Waseda University several dozen students worked, ate, and slept in a room the size of a two-car garage—easily the most densely populated, computerized, and mechanically equipped place on the planet. Slipping through the mob like a commuter on a rush-hour subway, Atsuo Takanishi showed us rat robots, a flute-playing robot, and a chewing robot made from a human skull. None were as startling as the Universal Dental Robot. It's used, Takanishi explained, to help women learn to chew again. What? Well, in Japan, the polite way for women to speak in public is a chirpy sing-song. For obvious reasons, some of them develop a syndrome in which they stop talking altogether. This causes their jaw muscles to atrophy and they ultimately cannot eat solid food. The dental robot consists of a mechanical bit inserted in the woman's mouth that gradually gets her back in the chew. Slackjawed, we watched a video of a patient using the robot. A drool bucket was part of the rig.

In an oddly ghoulish bit of dental R&D, Waseda University engineers have built a "jaw-robot" from a skull, some electronic circuitry, and an assembly of pulleys, wheels, and cables that act like muscle. Sensors measure the biting action of the jaw and the force of the chewing.

Like bees swarming over the cells of a honeycomb, the students at Waseda University's Takanishi Lab (see page 37) are moving back and forth between a dozen projects. Overwhelmed by the mechanical and human clutter, I initially can't pick out any one robot in particular until the sight of one stops me cold. My jaw drops; this robot has a jaw. Wedged between a rat cage for robot rat studies and a flute-playing robot that mimics the human windpipe, lips, and tongue is a real human skull on which is bolted an array of pulleys, cables, motors, and electronic circuitry.

I'm accustomed to thinking about robots in the sense that some day they might become human. I'm morbidly fascinated by this robot that once was human.

We discover that the robot is intended for research into mastication—chewing. In fact, the lab houses three dental-robot projects. In addition to the skull, there's a mouth-opening-and-closing robot (a joint project with Tokyo's Yamanashi Medical University) and a food-texture-measurement robot (a joint project with Wayo Women's University in Chiba). The skull robot is being developed by Hideaki Takanobu, a research associate in the mechanical-engineering department, and Atsuo Takanishi, for whom the lab is named. I ask Hideaki Takanobu to explain his mastication marvel.

Faith: This is one of the strangest robots I have ever seen. The skull is real, correct?

Hideaki Takanobu: Yes. Usually visitors don't care much about the mastication robot at first. But when we explain to them that this robot's skull is real, they are usually very surprised and are transfixed by it. I think we humans are interested in ourselves very much.

Why did you use a human skull?

In the beginning, we used a duralumin mandible [an aluminum-alloy jaw]. Then, in 1992, we used a plastic human-skull model. Now we use the real human skull that we got from Professor Maki of Showa University, who is our joint researcher. From building this system, I learned the complexity of the human mastication system. It is important to simulate this hardy motion that is so easy for people.

Why is it important? What do you expect to learn from using this machine?

Conventional research on human mastication used a theoretical model based on measured data about human jaw motions. Our robot can be used experimentally, because it can bite real foods with its human-shaped skull and humanlike muscles. In dental schools or hospitals, students may have difficulties if they only have figures or drawings to describe the human mastication system. By using the mastication robot, our 3-D and dynamic understanding is

greatly enhanced. Moreover, the quantitative jaw model being developed with this mastication robot will contribute to an infinite number of robot applications and dental/medical apparatuses.

How?

Conventional machines are able to simulate static biting situations or the jaw's position only. They can't simulate overall phenomena like the muscle motion, the sensory stimulus, or input from the central nervous system. Because the mastication robot mimics human bones, muscles, and sensory organs, it is a real simulator of the human mastication system.

I understand the scientific reasons for wanting to simulate chewing. But what would be the practical benefit of building such a robot?

Well, we have two objectives. The first is to develop a human-shaped robot to make an "engineered human model" [a complete theoretical model of the physical stresses and forces on a human being]. The second is to integrate the use of human-shaped robots—conventional mechanical, artificially engineered models—with new methodologies. In Japan, there is a disorder that affects many, many patients, especially young women, called TMD [Temporomandibular Joint Disorder]. It causes pain in the jaw and atrophy of the jaw muscles.

The Takanishi Lab has been jointly developing a mouth-opening-and-closing-robot series with Professors Ohnishi and Ohtsuki of Yamanashi Medical University since 1995. The robots integrate the human jaw model [developed by the chewing robot] and conventional robot technology to facilitate painless mouth-opening training on a patient's jaw joints. Before, this was not possible with the conventional training apparatus.

How do you expect that this kind of robot will be used in the future?

Our research will help identify suitable foods for elderly people—for everyone, actually—by analyzing the biting force during chewing. The quantitative parameters of foods are useful to develop or analyze real food.

And that's what you're doing with Professor Yanagisawa. How do you see your work advance the goal of building a humanoid robot?

At present, a combination of mechanical engineering and artificial intelligence is necessary for industrial robots. But now there are very few AI technologies that can be used in robots. Although mechanical engineering and AI are separate research fields at present, they will be combined into one field to create intelligent systems. Moreover, humanoid robotics is a large research field that integrates and develops artificial systems based not only on mechanical engineering and AI, but also on medicine, physiology, psychology, and other fields.

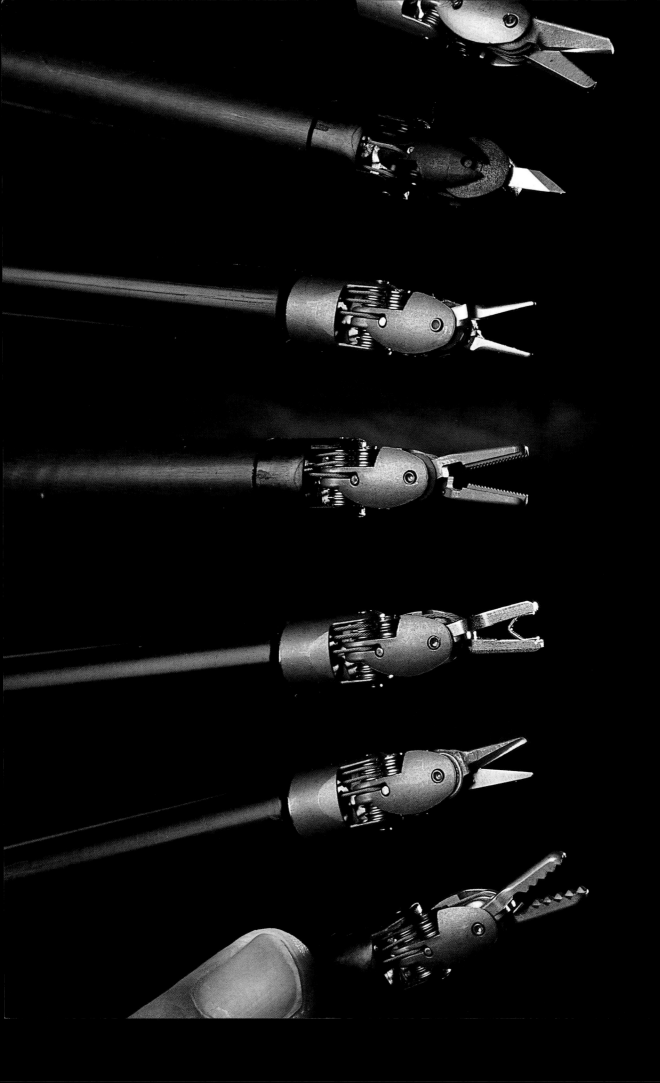

Kinder Cuts

Cardiac surgery traditionally begins with a full frontal assault. Surgeons saw through the breastbone, crack open the ribs, and then cut through muscle and tissue until the heart is exposed. Only then—after the patient has already suffered massive trauma—can the actual surgery begin.

In operations such as gall bladder removal, gynecological surgeries, and chest cavity exploration, surgeons are increasingly using less invasive techniques that involve inserting miniature instruments through small incisions in the body instead of breaking bones and slicing through muscle. For microsurgery, however, such as is needed in cardiac operations, the holes that let surgeons avoid sawing through a patient's chest are too small. Unless, of course, that hole can act as a conduit for an extremely steady, pencil-thin robotic appendage manipulated by an experienced cardiac surgeon.

Companies are racing to become the go-to supplier for minimally invasive robotic devices and techniques for cardiac surgery.

In one system, surgeons insert three remotely controlled, fingerlike metal probes—tiny surgical instruments, a light, and a fiber-optic camera—through small incisions in the patient's chest cavity. Seated at a console, the surgeon views a high-resolution, three-dimensional image of the surgical field from the camera. Using special joysticklike controls, the surgeon manipulates the robotic arm unit that in turn controls the miniature surgical instruments. Because the instruments are mounted on a kind of robot finger with a flexible mechanical wrist, it is capable of remarkably delicate movements—important in bypass surgery, which involves cutting and sewing delicate blood vessels on the surface of the heart.

Although this system has been approved for use in Europe, it is still considered experimental in the United States. Volkmar Falk, a cardiac surgeon at Germany's Leipzig Herzzentrum who is now a Stanford postdoctoral fellow, was a skeptic until recently. How, he asked, could doctors operate on a heart that they weren't touching with their own hands? These days, he says, he can answer his own questions.

Faith: Do you recall the very first time you sat down at the console of this system?
Volkmar Falk: Yes, I recall it very well because I had just come from a test run of another system on the same day. It was like the difference between a Volkswagon and a Mercedes.

In what ways were they different?
Both are tele-operated systems—that's the common ground. But in the way the systems are constructed there are many differences, a lot of them of a technical nature. This one has six degrees of freedom at the tip of the instrument, two more than the other system, and that allows you to transfer any motions naturally, from your hands onto the tips. It overcomes the limitations of conventional endoscopic instruments that have only four degrees of freedom.

The other system doesn't have a wrist kind of mechanism?
They're trying to build one, but it's not steerable from the console and it's not intuitive in the sense that it translates all the surgeon's hand motions online to the tip of the instrument.

Now that you've become accustomed to tele-operated surgery, is it hard to go back to conventional surgical procedures?
[Laughs.] Since I'm used to doing traditional surgery, that is still my home base.

It's not confusing to switch from one to the other?
I have to say that sometimes, for a break, it's nice to do open surgery.

To remember what it all looks like in the flesh?
Right. In a way, looking into a video screen for five hours is exhausting. But I believe we will get faster and faster on the system.

How fast can you get? Can this make surgery faster, as well as easier on the patient?
That's hard to know. A year ago if you'd have asked me, "Could you do this at all," I would have said, "No, it's impossible." Now we do coronary bypass grafting through three ports almost routinely. It takes a while, but I can see that we will come down in procedure times quite a bit. And in fact we've now done endoscopic beating-heart by-pass grafting surgery; it is difficult though!

So you don't have to stop the heart and put the patient on a heart-lung machine, which you have to do with conventional open-heart surgery. It sounds easier on the patient, but does it in actuality create fewer postoperative complications?
It's too early to say. We don't know. There will be different complications that we don't know of. To the question about avoiding the heart-lung machine—there are proven benefits of avoiding it, so off-pump surgery is one of the goals for modern coronary surgery. It's too early to say we have no complications or fewer complications; or more complications but less trauma, which might reduce the morbidity [the likelihood of health problems] from the procedure. But we know we have long-term good results which create strong arguments in favor of this therapy. At least that's my hope. Otherwise, I'll need a new profession soon [laughs].

What's the recovery rate as compared to normal surgery? Is it much faster?
I can't say right now, because the numbers are still too small. Certainly not longer. Some people recover real quickly, but in general it's too early to make

Robo Specs

Name: da Vinci™ Surgical System

Origin of name: Leonard da Vinci, who was a scientist and an artist—just like many great surgeons who must reconstruct anatomy post-disease or injury.

Purpose: To perform more precise, less invasive surgery.

Vision: 3-D stereo

Computers: Our own—based on 4 Shark processors

Funding: Mayfield Fund, Morgan Stanley, Sierra Ventures, other private investors

Project status: Complete— 11 systems installed (9 in Europe, 2 in the US)

Information from: Sheila Shah

Pencil-sized robotic surgical instruments allow heart surgeons to perform operations through a centimeter-long hole in the patient's chest. In a procedure that greatly lessens the need to cut into tissue—thus reducing postoperative pain and recovery time—doctors insert robotic instruments through minute "ports" in the body. Instead of hovering over the operating table, surgeons sit at a console a few feet—or, in theory, a continent—away, controlling the robo-scalpels with a pair of joysticklike grippers. Each tool has a patented EndoWrist™ mechanism that allows it to move with the dexterity and precision of the human hand. The whole ensemble—console, tools, and operating table—was developed by the Stanford Research Institute, a nonprofit R&D center created by Stanford University. The system was commercialized by Intuitive Surgical of Mountain View, Calif.; it now costs about $1 million. The da Vinci™ surgical system, as it is called, is now in use in Europe; although the da Vinci™ system has not yet been approved by the Food and Drug Administration, the first experimental remote-control bypass in the United States took place in September 1999.

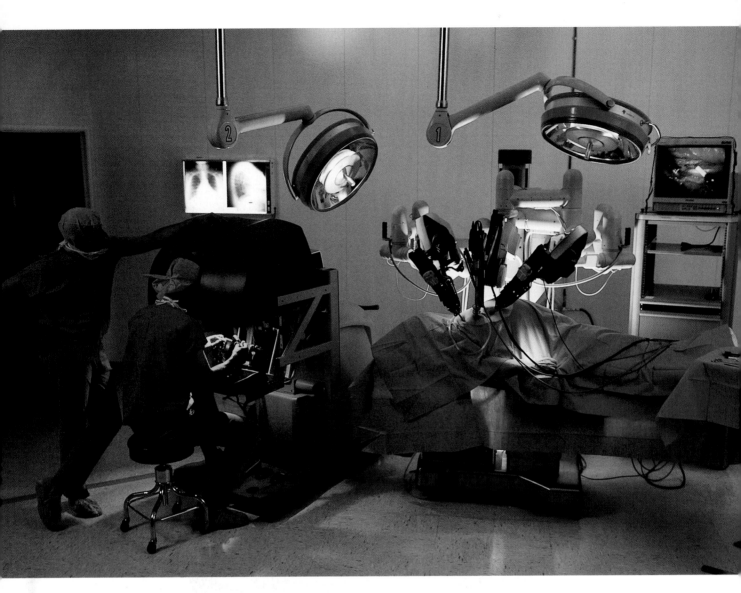

Burying his face in a 3-D viewing system, Volkmar Falk of the Leipzig Herzzentrum—Germany's most important cardiac center—explores the chest cavity of a cadaver with the da Vinci™ robotic surgical system. Thomas Krummel (above, standing), chief of surgery at Stanford University's teaching hospital, observes the procedure on a monitor displaying images from a pair of tiny cameras in one of the three "ports" Falk has cut into the cadaver. The other two ports are for the robotic surgical tools (previous pages). Falk sees an enhanced version of the same image: software merges optical data from the two cameras into a full stereoscopic picture of the operating site. Because Falk is working with a cadaver, the tools are not swathed in protective cloth, as is the case with his live patients in Europe.

statements. Eventually it will be quicker, I am sure. People will experience less pain because there are no incisions. The procedure time is somewhat longer, though, so it kind of levels off between those two right now. But we are on a fast learning curve.

You could also do simulation surgeries with this system, couldn't you?

The simulation could be exactly like an actual surgical procedure. You could pick tapes from an operation that has just been done. This will also revolutionize surgical training. We're thinking of coupling two of those consoles, like in driving school, and then you could have one be the instructor's and one be the trainee's.

The experienced surgeons would have an override console?

Yes. The way surgeons train now hasn't changed over the last hundred years. The problem is always that the surgeon who is doing the surgery has a different view of the procedure than the surgeon who is assisting. If it's a difficult case, usually what happens

is that the more experienced surgeon takes over. Sometimes the problem is not a big motion or move. You have the needle aligned in a certain direction and it is easier if you change it a little bit. But because it's hard to explain verbally, what happens is that everybody gets frustrated and the trainee doesn't end up doing the surgery. In this system everyone would have the same view and the supervising surgeon could redirect the needle and guide the stitch.

When I tried the system on the research cadaver, it didn't feel like there was any kind of feedback from the instruments to my fingers when the instruments touched the tissue.

There's actually very little. You don't feel those small tissue forces, but that can be compensated for by visual cues, which you have a lot of because of the magnification of the camera. What you will feel is bony resistances or if you have collisions between instruments, so anything that is a strong force you will feel. This alerts you that something needs to be changed.

You only get that kind of feedback when there's a problem?
If you look at this tissue here and push it around with this instrument [demonstrating on the cadaver], if you displace it by six millimeters—you need to push the tissue at least that much to feel the difference with your fingers. The sensors can't detect anything lower than that because the forces generated are too small. On the other hand, I can see a displacement of just one millimeter. My vision is much better at detecting spatial deviations than my tactile feedback.

When I looked through the headset there was amazing visual depth. I felt totally immersed in the image.
With the high-resolution image I have more visual clues than ever, so I don't really miss the tactile feedback too much.

Are older surgeons somewhat reluctant to learn this kind of procedure?
There is some prejudice by the older surgeons. They say, "I need to palpate, I need to feel." And of course that's helpful, but even when you use conventional surgical instruments, there is almost no tactile feedback in the range of what we need.

When you manipulate these tiny instruments, do they feel like scalpels or scissors to you?
Yes, but that's artificially generated in the handles. It's not the real force. It's kind of relative.

Is that something you have to learn to deal with or does it come pretty naturally with your other skills?
You have to learn it, but it's not like it takes days or weeks.

You were telling me about the operating rooms in Leipzig. The surgeon is not in the same room as the patient?
The idea was that the surgeon is separated from the patient anyway, so you might as well move him out of the room.

Can you communicate with the staff in the room where the patient is?
We have microphones. We have only selected information flow between the system surgeon and the table, and between the nurse and the surgeon. Everything else is eliminated.

But do you find yourself wondering about what you're not hearing?
No, I'm happy that I don't hear them anymore.

Really?
Because it's a new way of surgery and you want to be totally immersed in that image. Your brain can only process so much information. If you hear something—the brain has to process it. To concentrate, you try to shut off all those inputs you don't need, which leaves more space for the rest of the information. That's why we did that.

Intuitive Surgical's Sheila Shah also answers:
Faith: People are familiar with traditional laparo-
scopic procedures, in which doctors perform surgery in the abdomen by inserting a long tube—a cannula—with a special mechanism on the end that combines a little camera and surgical instrument. How is this procedure different?
Sheila Shah: With our system the surgeon using the console is visually immersed in the field. With standard laparoscopic procedures you are working with long instruments and the control is way up at the top. You're looking up at a 2-D screen and you're working backwards.

Backwards?
The instruments go through the trocar [a kind of collar for the cannula]. It's at a fixed position, and if you want to move the tip of an instrument right, you have to move the back end left and visa versa. So what we've

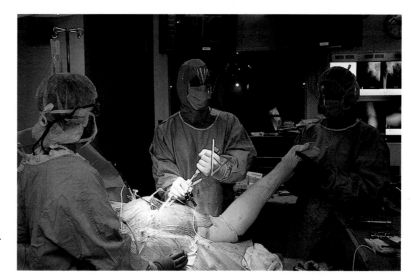

done is taken the controls out of the surgeon's hand and shrunk them down to the tip of the instruments only. If they make a movement to the right, the instrument goes right and so on. It's very intuitive. Every movement the surgeon makes at the console is simultaneously transferred to the tip of the instrument.

Does the fact that it's robotic make people nervous?
We are trying to call it computer-enhanced because it is more than just a robot. It's not something that you program and the robot does it by itself. All over the world that kind of automatic machinery is described as robots and we're trying to stay away from that. But we can't really stay away from it.

So are patients scared when they realize they're going to be operated on with a robot?
I have prospective patients calling me, saying, "Who can I talk to? Where can I have the surgery? Who can I call? I refuse to have my chest cracked open, I want to have the three small holes!"

They may be scared of robots, but they're more scared of open-heart surgery.

Peter's Notes

Impressed by the robo-surgery, Faith rented another cheesy old science-fiction video—we'd already seen *The Stepford Wives* in response to an earlier robot I'd photographed. In *Fantastic Voyage*, a miniaturized submarine full of scientists is injected into the bloodstream of a spy to zap an inoperable blood clot. Fast-forward to today, to what we had just seen—a heart surgeon took us on a tour inside a human body, using tiny, robotic instruments inserted through three small holes in the chest. Just like the movie, we explored the subcutaneal frontier. Except this was better—the view through the surgeon's goggles was in 3-D.

At an early-morning procedure at Shadyside Hospital in Pittsburgh, Anthony M. DiGioia (above, center) uses HipNav—a computerized navigation system he developed in collaboration with Carnegie Mellon's Center for Medical Robotics and Computer-Assisted Surgery—to replace the hip of a 50-year-old Pittsburgh man. Aligning the new hip properly, DiGioia explains, is necessary to avoid surgical complications. Here DiGioia, a former robotics student, uses the intraoperative guidance system and a simple "aim and shoot" interface to emplace the new hip. If the angle of insertion is not correct, the computer yelps an uncannily accurate rendition of Homer Simpson's "doh!"

Armed Man

Robo Specs

Name: Motion Control Utah Arm 2 with Motion Control Hand

Purpose: Rehabilitation of arm amputees

Creative inspiration: Human arm and hand

Length: 27.3 cm

Weight: about 1.3 kg, with Hand and cosmetic glove

Actuators: Basket wound motor with permanent magnet, 1 motor for each degree of freedom (i.e. elbow, hand, and wrist movements)

Load limit: 22.7 kg with elbow locked

Feedback sensors: Position, force

Frame composition: Composite of nylon with carbon fiber and fiberglass

Batteries: 12v Nickel Metal Hydride; rechargeable

Cost: $50,000 to $75,000

Funding agency: National Institute of Health

Project status: Ongoing

Information from: Harold Sears

Peter's Notes

Spending a few days with Bob Goodman at his home in eastern Oregon was a real treat. He lives in Halfway, named for its location between the towns of Robinette (now under water) and Cornucopia (now deserted). Quiet town, big sky, friendly people. In 1989, Bob was putting up a TV antenna when it came in contact with some high-voltage wires. His right arm was burned off and his leg was also burned where the electricity exited his body. The sparking was so bad that it started a grass fire which burned up a nearby trailer. His heart stopped, but when he fell off the ladder the shock of hitting the ground restarted it. He was taken to Portland by helicopter, where his arm was amputated. After skin grafts and reconstructive surgery on the stump, Bob went to Salt Lake City, where he was fitted with a myoelectric prosthesis at Iomed Inc., which is now Motion Control. With the insurance money from the accident, Bob was able to upgrade his living conditions to an 80-acre ranch outside of town. Today, he continues to run his ranch, do carpentry and odd jobs, and cook and clean for himself. When we met, he told me he had just begun skiing again.

Rancher and carpenter Bob Goodman can't remember the moment he was shocked by the high-tension wire while doing some work atop his neighbor's roof, but the ten years since have been a constant reminder. After his badly burned right arm was amputated above the elbow, he was fitted with a prosthetic limb in Salt Lake City. His new myoelectric prosthetic arm uses an electrical signal from his own muscle to control a motor that operates the realistic-looking hand. The myoelectric arm was harder for Goodman to learn to use than an ordinary prosthesis because he had to learn how to flex his muscles to send the proper signals to the arm and hand. Nonetheless, Goodman says the extra mobility is worth the trouble. He has replaced the first myoelectric arm with a newer, more advanced version.

Faith: Do you think of your prosthetic arm as part of you Bob?
Bob Goodman: I was amazed at how soon it really did start becoming a part of me and I didn't really have to concentrate on flexing my muscles to make my hand open and close. It just became second nature to open and close the hand and release the elbow. That's the hard part; you have to flex both muscle sites at the same time and then relax and then release the elbow. It took me awhile to get the muscles flexed at the right time.

How long were you without an arm before you actually got fitted?
I had a fair amount of surgery on what remained of my arm, so it was pretty sensitive to being fitted with an artificial arm. It was probably three or four months.

You told me that you can't imagine not having your prosthetic arm, but I would imagine there are things you would like to do that you still can't.
Yeah, and a lot of 'em are simple things. I really miss just a two-armed hug, right off the bat, you know, or dancing. I used to dance a lot. It's funny how I can farm and build houses, but then there's these little things that come up that are so simple, you would think, but they are difficult. Sometimes those Ziploc bags—it's hard to get the finger in there. You're using your teeth a lot on stuff like that.

I can't do that very well with my regular hands [laughs].
Well, that makes me feel better. Let's see—other things. Holding nails is no problem, and a lot of times for big jobs I use a nail gun. But little finish nails or little brads—that was really kind of a setback for me to deal with that. So, I did have to tool up quite a bit more. I used to love to hold a chisel—now I have just had to get some different tools to keep on woodworking the way I like to.

You have an updated version now, which has a long-lasting nickel-metal hydride battery. Did the battery in your last arm ever die?
It sure did. If I happened to have the elbow locked and I didn't know that the battery was getting low, then all of a sudden it would go completely dead, and I couldn't unlock the elbow. There's an override, but I had to take the forearm off, then take one of those little nine-volt batteries and do a jump to get a little spark going—

So you jump-started your elbow?
Yeah, and that released the elbow—that only happened to me once.

What were you doing when it happened?
I was just around the house, so it was real convenient to break it down, but I didn't have a nine-volt battery handy, so I took one out of the smoke alarm.

The arm must be pretty reliable if it only happened once, but that was quick thinking.
I'm glad I did think of it. I think there is going to be a manual override on the newer arms. There will be a little switch or button somewhere that you can push and it will release the elbow, so you can put a new battery in.

Is the arm heavy?
It is kind of heavy. I believe it weighs about the same as a live arm, but there is more weight distributed toward the hand, and I hear that's why a lot of people don't opt to wear myoelectric, or if they do it's just to go out to dinner or something, because of the weight of it.

What's the most surprising question you've been asked about your arm?
A couple of years ago I was giving a talk to an elementary school, it was a third or fourth grade class, and one boy asked, "In your dreams, do you have your artificial arm on or not?" It was a wonderful question and something I was paying attention to.

What did you tell them?
That for the first couple of years I was dreaming with two arms. Once in a while I was dreaming with an artificial arm, or no arm at all.

The way mass media sometimes portray technological breakthroughs, do they do prosthetics a disservice?
I think that's a big hindrance for an amputee—believing that they are going to receive a prosthesis that's a bit like something in a James Bond movie. Well, it isn't. With myoelectric, you've got to mentally flex the muscles you are working with. It would be really frustrating if you're going in there thinking right off that your hand is gonna be opening and closing and the elbow locking and unlocking very effortlessly. It's hard work in the beginning.

What realistic technological advance do you hope for?
I'm pretty excited about the idea that someday a hand could operate more like a robotic hand, with each finger having individual movement. The point of contact on the hand now is still pretty limited. It's just the thumb, and then the middle and pointer finger as a solid unit.

A rancher in Halfway, Ore., Bob Goodman (above) lost his arm below his elbow in a freak accident. Researchers at the University of Utah attached a myoelectric arm, which he controls by flexing the muscles in his arm that are still intact. Sensors on the inside of the prosthetic arm socket pick up the faint electrical signals from the muscles and amplify them to control the robot arm. In this way, Goodman can cook his dinner and do his chores (left), just as he did before the accident.

Prosthetic Aesthetic

Robo Specs

Name: Intelligent External Knee Prosthesis

Purpose: There is a tremendous need among above-the-knee amputees for a reasonably priced, intelligent knee prosthesis that adapts to speed and terrain variations.

Height: 15 cm

Weight: 1 kg

Sensors: Force, moment, angle

Frame composition: Aluminum, titanium, carbon composite

Batteries: Yes

External power: No

Person-hours to build robot: Approximately 23,000

Project status: Ongoing (plan to finish by the end of 2000)

Information from: Hugh Herr

Name: i-PASS (below)

Origin of name: Acronym for Integrative Physical Assist offering Seamless Service

Purpose: To reduce the difficulty of transferring elderly people from a traditional bed to the wheelchair and the heavy physical load this puts on caregivers.

Length: Can fit through a 26" doorway with room to spare.

Weight: 45 kg

Project status: Ongoing

Information from: Joe Spano

Driving with a joystick, MIT graduate student Joseph Spano (right) takes a spin in the ball-wheelchair he is helping to design. The chair, which uses spheres instead of wheels, automatically compensates for movement—if Spano reaches down, the chair responds by thrusting out its "wheels" to prevent him from toppling over. Such robotic autonomous-control technology will become more and more useful to the disabled—as Hugh Herr (far right) can testify. A double amputee, MIT Leg Lab researcher Herr is developing a robotic knee. Standard prosthetic joints cannot sense the forces acting on a human leg. But a robotic knee can sense and react to its environment, allowing amputees to walk through snow or on steep slopes now impassable for them.

For two months, in 1982, doctors tried to save climber Hugh Herr's frostbitten hands and legs after a harrowing wintertime rescue from New Hampshire's Mt. Washington. They saved the seventeen-year-old's hands but had to amputate both legs below the knee. This would be a life-changing moment for anyone, but for Herr, it was life-defining. Now thirty-five and a principal investigator at the Leg Lab of the Massachusetts Institute of Technology, Herr, who still climbs, is trying to help people who have even more severe injuries.

Hugh Herr: Specifically, the medical problem that now exists in lower-extremity prosthetics is that the mechanisms are primarily fixed, passive devices. They do not change with walking or running speed. They do not change with turning. Prosthetic and orthotic [bracing] devices are also passive, meaning that they cannot power a walking or running step—unlike a real leg, which pushes a person forward, you don't get thrust from the artificial joint.

Faith: How do you go about building a less passive artificial leg?

Our goal is to make the knee joint not only adaptive to changing speeds but fully adaptive—patient adaptive. What that means is that the limb-maker will be able to take the artificial knee out of the box from the manufacturer and bolt it to the prosthetic socket, the interface between the artificial leg and the human leg. And from there, perhaps with, at the most, a single input—say body weight—the leg will conduct experiments and adapt itself to the patient.

So it learns from the patient?

It continually adapts to what may be a common disturbance such as a patient changing his shoe. The system would always be looking for such a disturbance and its characteristics.

Or recognizing that a walking surface has changed?

Or a person is walking from a rigid surface such as cement to a surface such as grass. Now the prosthetic foot is swinging through grass, which slows it down. The system would say, "Ah ha! The leg has slowed down!" and it would reduce the resistance at the knee, so it could go to complete extension in the same amount of time as it did on the rigid surface.

For now, you're trying to figure out the best means to build a mechanical leg. But I understand you'd like ultimately to create something much more like living tissue.

To start, we will build machines with real, live muscle tissue—small machines, with simple, open circulatory systems designed to nourish the muscle tissue. Then, when that is mastered, we will scale up to a muscle that is as large as your gastrocnemius [the largest muscle in the calf of the leg], which can actually power an ankle.

Do we know enough about the human body to be able to create an artificial muscle that works just like a real muscle?

No, or it would be here today. What's interesting about these tissue-machine hybrids is that you don't have to understand how it works. You just have to get it to work. The understanding can come later. What's fun though is that this problem is right smack in the middle of several fields.

Anything robotic seems to be very multidisciplinary.

In the science world, the wonderful new things about these hybrid machines is that we're getting closer and closer to building an animal. Now we can play God. What happens when we make this particular tendon a different stiffness? We believe such-and-such is going to happen based on this theory of animal locomotory function [independent movement]. That's exciting.

We have robotic knees, robotic ankles, robotic interface. What's the connection between those things?

Once you have the individual modules, you are absolutely right, they should talk to one another. We're not going to develop a technology that enables lower extremity amputees to walk fluidly until the ankle and the knee talk to each other—it's pretty clear. That's also down the road. But we first have to develop the individual components and really have the machine designed.

Hearing about these hybrid mechanisms makes me wonder how much people will merge with technology and become machines.

I wouldn't say "become a machine," but as time marches on the machine and the human interface will become more and more intimate. On a positive note, what could come out of this is that we could all become visionaries and the technology around us could do the work of middle managers and what-not—leaving people free to think of good things, important things.

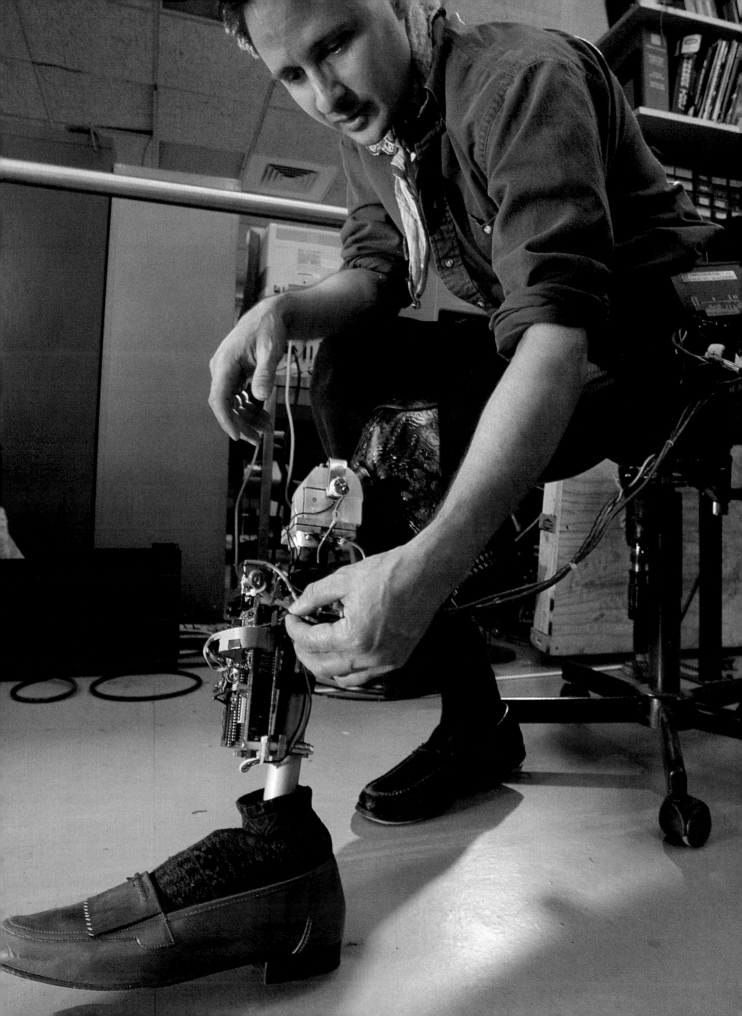

Step Children

"**W**hat's interesting about this model," Dan Paluska says of his robot mock-up, "is that it has no motors—only a couple of rubber bands near the hip and at the ankle to set the default position." Paluska, a graduate student in the MIT Leg Laboratory, makes the robot model walk by holding it at the hip and pushing forward. That's enough: the structure of the robot automatically translates the push into a walking motion. "When you watch an entire walking cycle," Paluska says, "you notice that the knee on the swinging leg does in fact bend and swing through on its own and sort of lands in the right place. That shows that you don't have to have everything preprogrammed."

Faith: Dan, this does not look very much like a robot.
Dan Paluska: This is only a foam prototype of the new 3-D biped robot that we're building.
Do you always build a foam model first?
It's a very common practice in mechanical engineering, because it's really quick and when you're working on something on a computer all the time you sort of lose track of how big things really are. I had all these computer-generated pictures of this robot, but after I built this model, everybody was saying, "Oh, my gosh, I had no idea that's what it looked like." Even people who understood came to a whole new level of understanding once they saw it. But actually, none of the original design remains in the current design of the robot.
What kinds of things did you change?
The actuators—sort of the muscles of the robot—changed so all these multiple tube structures in the legs are now single tube structures with actuators on the sides.

The whole idea is that the actuators have springs in them and the springs serve two purposes: they're sensors and shock absorbers. So when your leg impacts with the ground, which is inevitable in walking, then you have a little bit of springiness, so it doesn't hurt the structure of the robot. And that's why our tendons have springs.
What's your prognosis for progress?
I think this robot will do some neat things that other robots can't do, i.e., walk in an uncertain environment. And walk with a little bit more of a natural look than the other legged robots. Hopefully, it will be more efficient, because it takes advantage of natural dynamics. Robots like this will use less power than others.
And this is a lab-rat robot?
It will never walk outside. This robot will absolutely not have a use outside the lab because once this

thing is down it can't get up. It will be able to take small stairs and ramps, in theory. It's designed to be able to do that. I'm not sure it will.
You're a mechanical engineer—would you call yourself a roboticist?
I guess so, yeah. It's really odd to think that I design robots that walk. I guess none of the robots that I've designed have walked yet, so we'll see whether it's a fitting title.
What do you think about the increased interest in full-bodied humanoid robots? In Japan, they're in the midst of a nationwide project to build a humanoid robot platform. They call it the Grand Challenge.
It's really neat but I think it's too early.
What about Honda's P3 humanoid [see page 43]?
I think the Honda robot is incredible. It's really amazing. When it came out I was completely floored. Although the technology's wonderful and the work floors me, what really gets me the most is that a major corporation that has a bottom line and is not a research institute spent ten years making a robot that looks like a guy in a spacesuit [laughter].
Your own lab has generated some really interesting concepts. Marc Raibert's running robots were something never seen before.
Yes, he came up with these incredibly simple solutions that were really easy for anyone with even a minor engineering background to understand. And the machines themselves were really neat looking. And just the motion—when you have something with motion like that, that's so different, so completely different from any robotic arm that you've seen—I mean a robot doing a front flip is just, like, wow! My robot will never look that cool. I know that already [laughter].
Technically, the Leg Lab is part of MIT's AI Lab. How important is AI to your robot?
I'd say it's pretty low on my priority list, although there is this idea of the intelligent machine—that the mechanism does the computation for you—and that's sort of the idea that this swing leg is your natural motion and you don't have to control that. I think of that as an AI of sorts. When the thing is walking well and when everything's all reasoned out, I think the idea of this thing then learning to do things is really cool. I do like AI in that sense.
Is it very real?
I don't know. That's a good question. I'm in the AI Lab and I see all these people doing neural networks, genetic algorithms, blah blah blah, buzzword this, buzzword that. I really would like to get into these things, but they would all require *x* amount of time and since I've been spending all my time designing the robot, I don't have time for these other things. Hopefully, when the mechanical part is all done, I'll get a chance to write software and investigate.

Robo Specs

Name: Pinky

Origin of name: Robot's color

Purpose: To evaluate design possibilities for a 12 degree of freedom active humanoid bipedal walker.

Creative inspiration: Thin air and Marc Raibert's three-dimentional hopper robot and the human form.

Height: 1.3 m (.9 m legs)

Weight: A couple of pounds (~1 kg)

Vision: None

Sensors: None

Frame composition: Pink foam, wooden dowels

Cost: $50

Project status: Complete (working on successor robot)

Information from: Dan Paluska

Peter's Notes

I love to visit the Leg Lab at MIT. I always walk out of there with a much greater appreciation for my own seemingly effortless bipedal locomotion—the amount of complex engineering work and computer power it takes to make something walk is phenomenal. Hanging from the ceiling of the lab are some of founder Marc Raibert's early walking robots, which moved the field from slow, painstakingly computed, crab-on-Quaaludes stepping motions to the jaunty dynamics of spring-loaded kangaroo-type jumping. From robotic knees to Spring Flamingos to Dog on a Stick (a tethered quadruped on a treadmill, named after a corndog), the Leg Lab is chock-a-block with cool machines. Marc Raibert's current office overlooking Cambridge's seedy Central Square is filled with powerful computer workstations, his famous bipedal flipping robot, and a poster of Monty Python's John Cleese doing his famous shtick as the Minister of Silly Walks. Raibert's moves are still pretty inspiring, although now they are mostly virtual: his company is a leading provider of human motion simulation software.

Pinky (left, chaperoned by graduate student Dan Paluska) is the prototype of the next walking robot from the MIT Leg Lab. Established in 1980 by Marc Raibert, the Leg Lab was home to the first robots that mimicked human walking—swinging like an inverted pendulum from step to step. Famously, Raibert even built a robot that could flip itself in an aerial somersault and land on its feet.

Leaning back in his chair, graduate student Jerry Pratt controls Spring Flamingo, a walking robot at the MIT Leg Lab. A branch of MIT's renowned Artificial Intelligence Lab, the Leg Lab is home to researchers whose subjects run the gamut from improved artificial legs to robots that help scientists understand the complex dynamics of the human stride. Spring Flamingo, the Leg Lab's first walking robot with fully jointed ankles, can walk at a normal human pace and manage slopes of up to 15 degrees. Tethered to a slightly counterweighted boom that rotates around a pivot, the robot always walks in a circle in the lab. It has now walked about 10 miles, despite what even its creators admit is a propensity for stumbling. Sitting in the chair across from Pratt is the first model of Pinky (previous pages), Spring Flamingo's successor, which is intended to operate without a tether or boom.

Go Joe

Robo Specs

Name: The name of the robot is HelpMate, but each hospital names their robots. The robots at Danbury Hospital are named Rosie and Roscoe. Others are named ReX, Igorr, Obie, Hilda, Greta, Luke Hallwalker, Elvis, Lisa Marie, Mercedes, Bentley, Stosh, Rudy, Lil' Jeff, Petie, Jake, DT, Flash, Max, Cookie, Bart.

Origin of name: Robot helper.

Purpose: Joe Engelberger wanted to have a robot that directly worked with people. There was an apparent need in hospitals for extra help. Carrying items seemed to make sense.

Project status: Ongoing. Though much hasn't changed in the last 10 years, with the purchase of the HelpMate by Pyxis Corporation, we will surely be doing advancements.

Information from: Paul Cappa

Known as the founder of modern industrial robotics, 74-year-old Joseph Engelberger (far right) hitches a ride around his workplace on LabMate. A now-standard platform that companies use for a variety of autonomous robots, LabMate is manufactured by HelpMate, the Danbury, Conn. company Engelberger founded in 1984. (He sold the company in 1999 to Pyxis, a subsidiary of Cardinal Health, a health-maintenance organization in Ohio, but remains there as a consultant.) Engelberger's interest in robotics dates back to his days as a physics and engineering student. In the 1960s he founded Unimation, the first company that made large robots for automobile factories. Recently Engelberger has devoted more of his energies to making robots that can move about and interact with people—the focus of HelpMate. The task, he admits, is daunting. "You can make them foolproof," he says of his products. "But you can't make them damnfoolproof."

Whizzing around a hospital floor in Danbury, Conn., HelpMate (right) delivers patients' meal trays to hospital nurses. Now in healthcare facilities across the United States and in several other countries, the battery-operated, $100,000 HelpMate can transport food, drugs, and hospital wastes and even perform simple patient-care functions, such as guiding patients through hospital corridors. When sensors detect people blocking its way, it can steer around them—or, if that's not feasible, politely ask them to move out of the way.

Walking in the hospital hallway with her mother, a young patient points out a white box on wheels the size of a mini-refrigerator that is stationed outside the kitchen. "That's Rosie the robot," she says, and indeed it is—the robot is one of two Helpmate Robotics Inc. machines now assisting cafeteria staff with between-meal deliveries at Danbury Hospital in Danbury, Connecticut. Sharing the corridors with patients and staff, the autonomous courier robots use cameras, gyroscopes, infrared and ultrasonic sensors, and electronic mapping systems to guide themselves to where they need to go. In this structured environment mishaps are few, although Rosie would be lost outside the confines of its programmed environment.

Based in Danbury, Helpmate is the latest company created by entrepreneurial engineer Joe Engelberger, the industrial robotics pioneer. Engelberger got his start in the 1950s by recognizing after the Second World War that technology had progressed to the point where making robots was possible. But the key to his early success was recognizing the importance of a single patent. Called "program-article transfer" by its inventor, George Devol, the patent describes a mechanism by which machines could follow tapes of recorded movements to perform such repetitive jobs as picking up and moving parts on an assembly line. Because Devol-style machines could replay many different motions, they could behave as if they were learning new things. Engelberger and Devol launched a company, Unimation; the first Unimation robot, Unimate, was installed in an automobile plant in 1961. Today, thousands of industrial robots are in use.

After selling Unimation in the 1970s, Engelberger has focused his considerable abilities on the service sector. Helpmate robots are intended to help people perform routine tasks in offices and officelike environments, especially hospitals. Although the robots are now working in some 70 hospitals around the United States, no one knows better than Engelberger the vagaries of running a robotics concern, in which fortunes fluctuate with fashion, the economy, and the price of offshore labor. On the day we arrive at Helpmate's headquarters, Engelberger is announcing the sale of the service robot company to Pyxis Corporation, a subsidiary of Cardinal Health, Inc.

Faith: There are many roboticists who say that they always wanted to build robots, but you can't say that—when you began, robots didn't exist outside science fiction.

Joe Engelberger: It isn't as though I was dreaming of this as a kid. I went to Columbia University, some years after [the late Isaac] Asimov did—I didn't know him personally, but I certainly read all his science fiction. And when I got out of the Navy, I went to work for a company that built controls for nuclear power plants, jet engines, and those sort of sophisticated things.

The word robot was coined in 1920 but it took WWII to make a modern robot possible. You had to have servo technology, which was developed for gun aiming. Up to that point everything was done with adjustable stops and cams. And once servo technology came out, you needed digital logic, and you needed

Name: OmniMate

Origin of name: Omni from "omni-directional," mate from HelpMate, the company that built the platform.

Purpose: Demonstrate smooth motion and minimal odometry errors in an omnidirectional robot.

Height: 50 cm

Length: 183 cm

Width: 91 cm

Weight: about 200 kg

Sensors: 7 encoders, 1 fiberoptics gyro, 30 ultrasonic sensors

Batteries: 4 drive, 1 auxiliary

KLOC: 200

Funding agency: National Science Foundation, Department of Energy, UM Engineering Research Center for Reconfigurable Machining Systems

Project status: Finished

Information from: Johann Borenstein

Name: IRB 6400 (following pages)

Purpose: For manufacturing industries that use flexible robot-based automation.

Height: 2.24 m

Length: 1.9 m

Weight: 2,050 kg

Sensors: Position

External power: 200 to 600 V AC

Cost: ~$90,000

Information from: Fred "the Body" Barker

I love big auto factories. I wouldn't want to work in one—I'm too restless to stay in one building. But observing the process is fascinating—like watching an ant colony in which all the workers use very sophisticated, powerful tools to build an incredibly complex, mechanical beetle, one beetle a minute, 24 hours a day, 7 days a week. The first truly automated car factory I photographed was the Fiat factory in Italy, in 1983. There, in a system called "Robogate," robotic carriers quietly followed guide wires in the concrete floor as they moved engines from station to station. Borenstein's OmniMate uses precise measurements of its own movement to eliminate the guidewires. Like a car dealer proudly demoing a new model, he did an incredible job of parallel parking in his lab.

solid-state electronics because the first computers could fill a large room and still not be nearly as smart as what we have today.

Technological advances during World War II were the catalyst?

Yes. I was in aerospace control and that was the perfect background to do something like this. I met George Devol and I looked at his patent and said, "Geez, you know, you could call this a robot."

We got together after an evening of carousing in 1956 and I took him into the company I was working for. Then, a couple of years later, I formed my own company. It was still aerospace and nuclear controls; the robot was a sideline—the toy division part of it. We didn't have much money but we built the prototype.

That was Unimate?

Yes.

How hard was it to get people in general interested in the idea? Or, perhaps even more important, investors?

No one believed. Couldn't get money. To get Unimation started, I visited 47 different companies—all the logical ones like General Electric and Westinghouse. No interest. No money at all. We got the money from the Pullman Railroad Company. And why was that? All because the chairman had nothing to do [laughs].

What he was doing was fighting antitrust. The Pullman Company used to own the cars themselves and they built the cars. And the government said, "You can't do both." So eventually they sold the cars, which was the smart thing to do because the Pullman cars are oddities today. And they kept the manufacturing business. So here's the guy that's been fighting antitrust for seven years. I come in and I say I want $3 million for 51 percent of my company. Well, he's a big enough shot in that company that even if it's stupid he's not going to lose his job, right? So he puts the $3 million in and we start. It wasn't the rational way you'd expect something like this to start.

This was in 1961?

We put our first industrial robot out in '61 in a General Motors Plant in New Jersey. In '62 was when we got the money. And this cost thing—there are two ways to go about technology. One is to cream the market. Get high prices for a few things. And the other is to price it, to begin with, where it belongs in the marketplace. That gives you a bigger market, but now you've got to pay all the up-front costs. And when we started Unimation, we sold the robot for $35,000. But it cost us $60,000 to build, which doesn't seem like a very good idea [laughs].

Not for very long, anyway.

What Pullman did was hire a major consulting firm, and it said, "If you can get this up to a hundred of these a month, you can build them for $8,500 apiece." So you have to do some long-range thinking. You have to believe there's a market, get it up to the

right size, and then we know the economies of scale. *There was a market although it's never yet been really big. You say that the problem of getting more robots into use is more about money than technology?*

When no one has them, no one needs them. I tell people, "No one needs a robot. There's nothing a robot can do that a willing human worker doesn't do better." The thing is, there aren't many willing workers.

I was watching you flip through our photographs of robots around the world. What were you thinking?

I'm probably a bit controversial about what I think most of this stuff amounts to. For example, the Honda robot climbing the stairs [see page 44]—it's the silliest thing I've ever seen. Magnificent engineering, by the way—it's hard to do what they did. But for what reason? Its upper body doesn't do anything. It doesn't interface with the world. They triangulate to a spot on the floor. Our navigation is so much further advanced than their navigation. So here it is, it weighs 200 kg. Where's the battery? It's in the shoulders. And they have to figure out some way that when the power goes down the robot drops just straight down instead of falling over.

They kept it secret for 10 years. Nobody knew they were doing it from 1986 to 1996.

They should have let somebody know—it would have saved them a lot of money.

They had it on a stage on my last trip to Japan and the stage was pretty thin—plywood, I guess. It was like Frankenstein—*woompf*. It was laughable. With a wheeled vehicle, you can put the heaviest part near the ground. When you want to have two hands to pick somebody up, you need stability. You can't get it with that kind of design. They really have this tremendous urge to make it anthropomorphic in appearance. Everyone's doing that in Japan.

They have a long history of benevolent robots like Mighty Atom [see page 197]—the cartoon robot that was broadcast here in the 1960s as Astro Boy.

Their robots have got to be anthropomorphic. I don't like that.

Don't you think that people are naturally more interested in interacting with humanoid robots?

Yeah, maybe, but money is such an important factor. It doesn't cost much to have a contest where robots play soccer. The kids have a lot of fun; the advisors have a lot of fun. It's just like the Soapbox Derby. And it's limited in the budget. But you can't really play the robot game cheap. The real robot game isn't played that way. The fun part of it is. And they can continue with the fun.

While you do the serious stuff. It sounds like you don't think anything is learned from the playful, fun approach.

I'm sure that the people who do it learn something, but I don't learn anything. I'm waiting for one of those fun projects to ever contribute to the art.

What do you think about the Sony AIBO?

Have you seen the *New Yorker* cartoon? There's a
hydrant and the AIBO is there with his leg up. Nuts
and bolts are coming out. I'm going to use that in
my next talk in Japan [laughs]. It will be chock full of
Sony executives. They'll love it. If you look at it,
that's where the money has been made in robots.

Entertainment?

Yeah. Sure.

*You seem pretty negative about humanoid robots
in general.*

I say it's possible to make a humanoid robot but it
doesn't have to be a Stepford wife [laughs]. It's got to
get through the doorways that we go through and
it's got to handle our tools and our data, and live in
our environment. If you have that, it doesn't have to
look like one of us. And living in a human environment is tough enough as it is, by the way.

*So it's not just the way humanoid robots look that
bothers you?*

So many of these researchers want the robot to
develop intelligence itself. It's like somehow they

want it to be a baby who's going to experience the
world and as it experiences the world it's going to get
smarter and smarter. And I say, what's wrong with *a
priori* knowledge? Why not? If someone comes into
my plant and he's a new employee, I don't say, "See if
you can find the bathroom, you silly fool." I'm going
to tell him where it is [laughs].

*Rather than go through the trouble of teaching and
learning, you'd rather simply program the robot to
know what it is supposed to do.*

A priori knowledge is great! And you can load so
much into the robot. Memory is cheap. There are
still plenty of surprises after you have *a priori* knowledge. We have *a priori* knowledge in Helpmate for
hospitals. It knows every department, it knows how
to get up and down in elevators. But there are always
surprises—somebody left an X-ray machine in the
way and the robot's got to find a way around it.

*You are called the Father of Industrial Robotics. How
does that sit with you?*

All I can say is I was there first and I'm still here!

A "smart" pallet that can move in
any direction, OmniMate was
designed by Johann Borenstein
(above), a research scientists at the
University of Michigan. Like the
HelpMate hospital delivery robot
(previous pages), OmniMate sits
on robotic platforms called
LabMates. Although earlier robot
pallets had to move along cables
buried in the floor, OmniMate can
track its own location by measuring
its movements precisely. Once it
has gone to any destination, it can
remember its path and go there
again, which allows it to shuttle
between any number of locations
on a warehouse floor. Borenstein is
in the process of putting his robot
on the market.

Working around the clock without significant human input, industrial welding robots like these Swedish-made IRB 6400's build Sierra pickup truck bodies in the General Motors Truck and Coach Plant in East Pontiac, Mich. In the plant's body shop, 300-odd robots work to produce 73 truck bodies an hour—a pace that human beings cannot match. Working with uncanny speed and surprising quiet, robots are increasingly moving human workers from exhausting, dangerous, and repetitive tasks to more intellectually rewarding jobs. This workplace—which once contained many human laborers—is an example: the sparks are so intense that the area must be draped with plastic sheeting. Actually, the sparking in this photograph is unusually dramatic—it's a sign that the welding tips are wearing down.

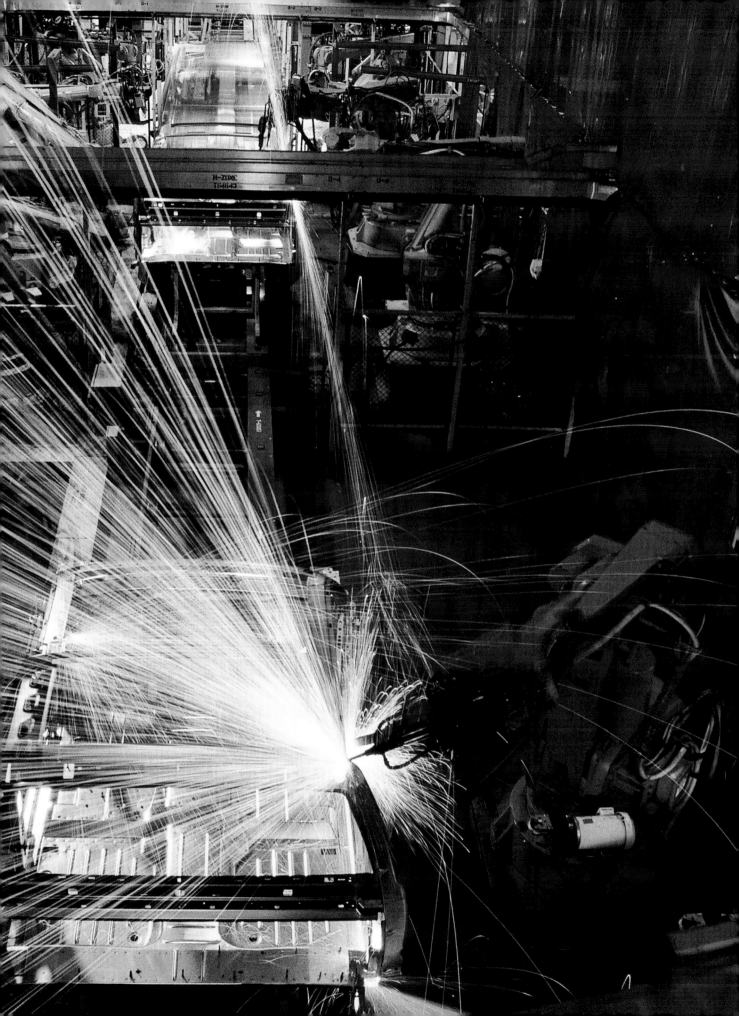

Ramp Up

There is not a human-shaped robot to be seen in the Hirose-Yoneda laboratory at Tokyo Institute of Technology. Instead, there are the generations of snakelike robots, quadruped walkers, wall climbers, stair climbers, and roller walkers that Shigeo Hirose and his graduate students have built during the last twenty-five years.

Hirose believes the human-shaped robot is impractical. "I showed a clip of the movie *Total Recall*," he says about a recent academic lecture. "Schwarzenegger is in a future city. He gets into a taxi and there is a humanoid driving it." Hirose used the clip as an illustration of what he considers foolishness—the intelligent automobile engineer in the future would imbed the driving intelligence into the automobile rather than

Robo Specs

Name: Titan VII

Origin of name: Acronym of Tokyo Institute of Technology No Aruku Norimono (in Japanese, TITech Walking Vehicle)

Purpose: Off-road vehicle

Creative inspiration: Observation of spiders

Project status: The entire series took 24 years, 40 students.

Information from: Shigeo Hirose

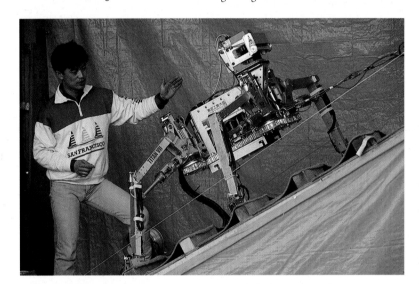

Although the Titan VII climbing robot (above and right) has only four legs, its designers drew their inspiration from spiders, which have exceptional climbing skills. Built by Hideyuki Tsukagoshi, a research associate in the Tokyo laboratory of Shigeo Hirose (see page 89), the machine is intended to be a mobile construction platform on steep slopes—the banks of highways and railroads, for example. (In safety-conscious Japan, such slopes are almost automatically regarded as dangerous, although they would not necessarily be treated that way in other countries.) At present, the Titan VII needs cables to walk up any hill—as shown by its progress up the 30-degree slope in this test facility in a garagelike lab at the Tokyo Institute of Technology. Ultimately, Hirose believes, the robot will be able to mount a 70-degree slope, although it may always require supporting cables.

clog up the front seat with a humanoid robot. Following Hirose's views, Hideyuki Tsukagoshi, a postdoctoral fellow in the lab, has built Titan VII, the latest in a series of four-legged robots meant to work in construction in Japan's many mountainous areas. It is specifically intended to climb steep inclines.

Faith: Why do you need a robot to do that?
Hideyuki Tsukagoshi: After cutting through mountains to construct railways and highways, workers are suspended by ropes to engage in tasks like setting ferroconcrete frames. But this is dangerous. Instead of workers, robots should climb the hills, spray the concrete and drill while stepping over ferroconcrete frames that have not yet hardened.
So it's better to have robots do this dangerous job. Can't other robots do it? What's special about this one?
Robots with wheels and robots with tank treads are simpler to control than robots with legs, but they cannot step over obstacles like ferroconcrete frames. Neither can they form the stable postures necessary on slopes by moving their centers of gravity. Only

walking-type robots can do this. Four is the minimum number of legs to do these actions statically. *They have to have four legs to balance themselves properly? I guess that's why people fall—why these jobs are dangerous.*

This quadruped walking robot has strong points from the standpoint of both hardware and software. In hardware, the length of each leg can be adjustable so the robot can adapt itself to the rough terrain. This is especially good for climbing a slope—it can shrink the legs on the uphill side and extend them on the downhill side. This enables the robot to walk more stably. But on the steepest slopes this strategy is not enough, so two cables tow the robot from the top of the hill. Controlling the tension of wires helps the robot climb more stably. Towed by wires, Titan VII could climb on a 70-degree slope. The software swings the body in zigzag fashion when it climbs. The center of gravity is always controlled to be situated in a stable position and the foot placement can be selected to step over obstacles.
Do you expect that this robot will actually be used in real situations?
Yes, I do. The construction example I just discussed is one application—especially under severe conditions that wheeled machines couldn't negotiate.
I know you of course built on earlier ideas, but what kinds of other ideas did you incorporate in this robot's design?
When climbing slopes, elephants fold their uphill legs. Looking at a film of this was very helpful in designing the robot.
What have you learned?
How clever humans are! They can climb slopes easily, if the slopes are not so steep. It took a long time for the robot to do the same action. To improve the robot, I think the feet must be designed much more intelligently.
Your robot is a mixture of mechanical engineering and computer science. Do you think mixing the two is necessary to build good robots?
Yes, I do. Especially from the aspect of learning. And we must study not only mechanics and computer science but also electronics, biology, human engineering—you name it!
Your former advisor, Professor Hirose, strongly believes that building a humanoid robot is a waste of time now. What is your opinion?
If the humanoid robot is developed to contribute to some specialized work, I think it is a proper project. If an accident happens at a nuclear power plant, a human-style robot could be helpful in the repairs. This is because the plumbing in nuclear power plants is so complicated that other types of robots would find it hard to access the necessary areas.
What are you working on now?
My main work at present is to develop practical robots for rescue operations. For instance, a self-propelled hose to get into the rubble, carry water, and rescue victims after a big earthquake.

Gas 'n' Go

It doesn't take a team of scientists to know when the sewer is backed up but it may take a team to figure out how to fix the problem. In a ten-year project, Frank Kirchner and his group at the Institute for Autonomous Intelligent Systems at the German National Center for Information Technology (GMD) are trying to build a robot to do the dirty work. As in his other projects (see page 113), Kirchner is focusing on robots that can work in the infinitely variable conditions of the real world, not just in the carefully controlled environment of the factory floor. "I read somewhere, and I believe it," he says, "that in times of change, it is the learners who will inherit the earth, while the learned will find themselves beautifully equipped for a world that no longer exists."

Faith: What can a sewer-inspection robot do that a person can't?
Frank Kirchner: Crawl into very narrow pipes. In Germany, it is forbidden by law to send people into pipes with a diameter that is less than one meter. But most of the 400,000 kilometers of sewage pipes in Germany are in the fifty-to-fifteen centimeter range. And another law forces the cities to inspect these pipes for cracks regularly, because leaking sewage causes environmental problems.

Hence a robot sewer inspector. How "smart" is the sewerbot?
As smart as we could make it. This means that from the current and past sensor readings, the robot is able to derive an action to perform. The robot can optimize its choice of actions during the interaction with the environment. We call this a learning robot, a system that increases its level of performance from experience. If any *thing* can be called smart, I think a system that learns is the smartest, as it is able to adapt to change.

How will these machines be used?
The robots will actually inhabit this sewer system and perform their tasks. In normal times, they inspect and maintain. They may form together to build little teams of robots that do a job. One robot brings the tools, the other robot brings the energy, and another robot may be videotaping the repairs for documentation.

Can they get lost?
Yes, we are dealing with a chain of ambiguous processes. Sensor data collection is not 100 percent accurate because of noise and failure; algorithmic processing of ambiguous data inherits nondeterminism, thus faulty decisions; and actions to be carried out in the real world are only partially defined. You can never be sure that you actually did what you intended to do.

Will the average sanitation worker be able to operate the robot?
Yes, it is as easy as making a phone call because the robot is autonomous.

A dirty, slippery sewer pipe wouldn't seem to be an optimum environment for a sensitive piece of electronic equipment.
The goal of my team is to build real-world robots. I believe that only robots that show sustained real-world behavior can be a basis for the study of cognition and phenomena like intelligence. By "sustained real-world behavior" I mean systems [robots] that actually live out there. They have to be built in a way that enables them to inhabit an environmental niche. Only in such systems can intelligent behavior emerge as a consequence of the machine's need to survive

Robo Specs

Name: Kurt I

Origin of name: German abbreviation for "sewer inspection robot"

Purpose: Inspecting sewers

Creative inspiration: The constraints of the environment.

Height: 26 cm

Length: 32 cm

Weight: 6 kg

Vision: Stereo camera

Sensors: Ultrasound, tilt

Frame composition: Carbon fiber

Batteries: 24 V 6 A

External power: Optional

KLOC: Approximately 1

Cost: $30,000 (components only)

Project status: Ongoing

Information from: Frank Kirchner

Kurt I (far left), a 32-cm-long robot, crawls through a simulated sewer network on the grounds of the Gesellschaft für Mathematik und Datenverabeitung—Forschungszentrum Informationstechnik GmbH (GMD), a government-owned R&D center outside Bonn, Germany. Every ten years, Germany's 400,000 kilometers of sewers must be inspected—at a cost of $9 per meter. Today, vehicles tethered to long data cables explore remote parts of the system. Because the cables restrict the vehicle's mobility and range, GMD engineers have built Kurt I, which crawls through sewers itself. To pilot itself, the robot—or, rather, its successor model, Kurt II—will use two low-power lasers to beam a checkerboardlike grid into its path. When the gridlines curve, indicating a bend or intersection in the pipe ahead, Kurt II will match the curves against a digital map in its "brain" and pilot itself to its destination.

Filling up a specially adapted Mercedes, the gas-bot (left) at the Institut Produktionstechnik und Automatisierung (IPA), a government-industry research center in Stuttgart, is intended for a time in the future when automobiles run on hydrogen. Hydrogen is an environmentally sound fuel—its main effluent is water. But it is also so explosive that robots may end up topping off people's tanks. A somewhat similar system for dispensing ordinary gasoline is currently being test-marketed by Shell in the American Midwest.

Robot Dreams

From the ranks of the post–World War II children who grew up on a steady diet of the exploits of beloved comic-book hero Tetsuwan Atomu, or Mighty Atom (also called Astro Boy), have sprung Japan's robot builders. Created in the 1950s by legendary animator Osumu Tezuka, the peace-loving boy robot had a nuclear-powered heart, rockets in his feet, and a computer brain. He fought society's monsters—real and imagined—and in doing so became the Japanese poster boy for mechanical goodness.

Among the roboticists inspired by Tetsuwan Atomu is Norio Kodaira, who designs and builds factory robots at Mitsubishi's Nagoya factory. Unlike Mighty Atom, Kodaira's machines can't walk around or interact with people. Instead, they belong to the robotic silent majority on factory floors: pick-and-placers, welders, clean-room workers, pipe benders, braziers, palletizers (machines that put boxes on shipping pallets). They are powerfully dangerous and kept isolated from humans—each in its own private space for as far as its arm can reach.

Looking at Peter's photographs of the other robots in this book, Kodaira says, rather apologetically, that his are not as interesting. Whereas laboratory robots have cute names and hybrid histories, his machines have bar codes and identification numbers. No smiling faces, no waving hands. "Just speed and consistency," says Kodaira. "That's what our customers want."

Faith: What kind of robots do you build?
Norio Kodaira: We have four major applications for industrial robots. One is a clean-room robot. The second is for logistics: palletizing, casing, and packing. The third application is micro-assembly. And the fourth is the small-sized general-purpose robot. Our mission is high speed and high accuracy.
It's a good mission for almost anything. Industrial robots are the workhorses of robotics. Still, what did you think when Honda [see page 42] announced that it had developed a walking humanoid robot?
Kodaira: Oh, the Honda robot is the dream of the robot engineer, so of course it is for me. Mitsubishi cannot spend any money for dreams because we sell industrial robots and the competition is very tough.
There is quite a bit of difference between the robots you build and the Honda robot.
If I had seen the Honda robot when I was a student, then I would have gone to Honda and gotten a job, but I'm in Mitsubishi now [laughs].
Could a humanoid robot like that be an industrial robot?
There is a very big gap between them. The Honda-type robot is not so useful for industrial applications because it is so expensive. The type of robot I have here is one million yen [$10,000 US] for a general-purpose, articulated robot arm; I think that the Honda robot is more than one hundred million yen [$1 million US].
It's a very expensive robot because of the research and time investment, but if the price came down, do you think then that it would have benefit for industry?
Factory owners don't want robots; they want working machines. A robot is only one of the choices for this. The factory owner can choose to use a robot or a special machine or human beings—whichever is cheapest. It's a matter of cost—if the human labor is cheaper than robotic, they'll go with human labor.
Have you always been interested in robots?
I went to the Tokyo Institute of Technology because the famous robotics professor Mori Masahiro taught there. I wanted to study with him so I went there.
I just read his [1982] book Buddha in the Robot, *which I thought was fascinating. It's still in print. Was it exciting for you to study with him?*
Of course. I am very happy about being in the robot business. I've wanted to be in the robot business since I was a high-school boy. My dream was to make Tetsuwan Atomu (Mighty Atom).
You and almost every other roboticist in Japan. Deep down, the dream is to build Tetsuwan Atomu.
Of course.

Industrial-robot designer Norio Kodaira (below) of Mitsubishi smiles proudly behind his Melfa EN, a robot arm that moves with incredible speed and dexterity to assemble pieces, drill holes, make chips—or just about any repetitive task that needs to be done quickly and precisely. Like many Japanese roboticists, Kodaira was inspired as a child by Tetsuwan Atomu (Astro Boy), a popular Japanese cartoon about a futuristic robot boy who helps human beings (right, a 15-centimeter Astro Boy action figure). Astro Boy, drawn in the 1950's, will soon be the star of a major motion picture. In the story line, his birthdate is in April of 2003.

Serious *fun*

Christian Ristow's bulldozer-tracked, raptor-clawed robot, called Subjugator, fires its flame thrower during a test run for his apocalyptic show of mechanical mayhem at the Burning Man Festival in Nevada's Black Rock Desert. A former Columbia University architecture student who is now an artist in Los Angeles, Ristow stages mechanical performances in which his constructions fight each other and destroy designated sacrificial targets. With typical bravado, he called his Burning Man show, "The Final Battle of the Twentieth Century Between Man and Machine for Ultimate Supremacy on the Earth."

Chop Till You Drop

Robo Specs

Name: Pretty Hate Machine

Origin of name: It was originally going to be called Mauler or something, but when the parts came back from anodizing they were just oh so pretty. So I borrowed P.H.M. from Nine Inch Nails.

Purpose: To beat up other robots in a free-for-all competition.

Creative inspiration: I just wanted to build a walking robot with a lot of rip-saws!

Height: 35 cm

Length: 1 m

Weight: 100 lbs.—wet (45 kg)

Vision: None—tele-operated

Sensors: None

Frame composition: Steel rod, aluminum cross ties

Batteries: 24 V

Engines: 2 E&J RAE wheelchair motors; 2 Weed Whacker 2-stroke gas engines

Project status: Complete—I don't care about creating highly innovative robotic solutions to modern-day problems. I don't care about unique approaches to implementing robot systems in industry. I just want to make cool robots.

Slogan on website: My robot can beat up your robot.

Information from: Christian S. Carlberg

———————

Fearsome sawblades spinning, Pretty Hate Machine (right) menaces the competition at Robot Wars, a two-day festival of mechanical destruction at San Francisco's Fort Mason Center. Organized by Marc Thorpe, a former Industrial Light and Magic model builder, the cybernetic slugfest spawned a six-week BBC-TV series and many similar events. Pretty Hate Machine is a middleweight-class machine; two wheelchair motors power a Rube Goldberg assembly of rods, rubber belts and saw blades. In battle, it dances toward its opponent on raquetball-tipped aluminum legs, its Weed Whacker blades whining menacingly. Contact abruptly changes the pitch of the sound to a metal shriek. A real crowd-pleaser, Pretty Hate Machine was one of the few walking robots in a competition dominated by wheeled or tracked machines. In a techno-robotic rendition of *Texas Chainsaw Massacre*, it tore up opponents Ziggy and Webster, then lost its third fight to Defiant by a judge's decision.

One of the strongest forces pushing technology is the apparently inexhaustible human desire for amusement. Children playing video games drove the evolution of computer hardware in much the same way that adults E-mailing dirty jokes helped foster the spread of the Internet. Roboticists have taken note of this phenomenon. Over and over, they tell us that the way robots will spread into society is through entertainment. As MIT Media Lab researcher Michael Hawley puts it, "Toys will be the Trojan horse that sneaks robotics into the home."

It's always been this way. In the seventeenth century, chess-playing automata dazzled the courts of Europe; harbingers of the industrial revolution, they helped accustom people to machinery. Not too much later, wooden tea-serving figurines—clockwork devices that walked toward people, nodding their heads and carrying cups of tea—enthralled the Japanese elite. There, too, automata were the advance scouts for the technological invasion that would transform the nation in the next century.

The importance of entertainment has reached an entirely new level in the developed world today. Not only have the mass media brought Hollywood movies and Nintendo to every living room—an onslaught amplified by the Internet—but the line between work and play is becoming ever more blurred. Workers in Atlanta play Net games on office time, salarymen in Tokyo download spreadsheets during vacation, and everyone trades stocks online just before turning in. When entertainment accustoms us to technology today, that technology infiltrates our lives faster and more thoroughly than ever before.

This process—the domestication of technology—would seem to face special obstacles in robotics. Though chess-playing automata were greeted with enthusiasm, many images of robots and robotic machinery are far more negative in Western culture: Frankenstein's monster, golems, the intelligently malicious machinery that swallows up Charlie Chaplin in *Modern Times*. The loyal C3PO and R2D2 in *Star Wars* are more than matched by the shapeshifting metal killer in *Terminator 2*, the treacherous android in *Alien*, or Yul Brynner's lethal animatronic gunslinger in *Westworld*. Robots, the academics say, embody people's fears of technology; clad in shiny, impervious steel, they are terrifying emblems of hubris, of the dangers of playing God by making living creatures.

Despite this cultural baggage, the robotic diversions we saw were vastly popular. All those images of Frankenstein notwithstanding, the entertainment industry knew how to market robots in a way that appealed to the customer. Indeed, one of the most crowd-pleasing uses of robotics actually took advantage of their fearsome reputation.

What could be more exciting than watching powerful metal creatures tear each other to pieces? That's what the organizers of Robot Wars thought, anyway. The last Robot Wars competition in 1997—before the legal haggling of the organizers shut it down—drew six thousand screaming fans to the Herbst Pavilion in San Francisco to watch dozens of largely homemade robots battle to the end. Gladiator-style, the robots dealt each other crushing blows. The winner was the last robot still moving around.

Robot battles are hyped-up tragicomedies, good

versus evil, sleek versus dumpy, bohemian versus classy all rolled together and radio-controlled. Spectators cry for blood and aren't disappointed. Strapped to the front of one mechanical contestant, Party Dress Barbie faces down the threat of an oncoming robot death machine—only to be decapitated within the first few seconds of the match.

In some ways the beheading of Barbie is exceptional; the robots often have a tough time getting started, much less doing any injury to each other. In fact, in the first matches, a lot of the damage is self-inflicted as robot operators learn what their robots can and cannot do. But, once the mayhem gets underway, the harm comes from many sources,

including the built-in dangers of the arena, — "stingers" that drop suddenly from the sky to skewer unwary robots.

The robots are built by hobbyists, engineers, special-effects wizards, and, sometimes, children and their parents; some builders spend the better part of a year working on their robots. Once they arrive at Robot Wars, the machines are divided into weight classes ranging from featherweight to heavyweight.

As might be expected, most of the children make small robots, which they enter in the featherweight and lightweight classes (along with a lot of adults). The only child-controlled middleweight is

Peter's Notes

Watching two nights of mechanical mayhem at Robot Wars was cathartic, but not enough to satisfy my gadget-bashing proclivities. When we got home, we took all the broken toasters, radios, microwave ovens, computer monitors, and kids' video games clogging up our garage, arranged them aesthetically in our cactus garden—and set them on fire all at once. The photograph of the conflagration became our next year's greeting card—a toast to technology, neo-Luddite style.

Ringed by six-foot sheets of bul-
let-proof glass and a sellout
crowd, radio-controlled gladiators
battle to the mechanical death.
At Robot Wars, a two-day-long
competition in San Francisco, the
crowd roars to the near-constant
shriek of metal, the crash of flying parts, and the thunderous
beat of techno music. After a
series of one-on-one matches,
losers and winners alike duke it
out in a final death-match called
a Melee. In this Melee, the
13-foot Snake (left) curls to use
its drill-bit tail on its hapless victim, a tracked vehicle; meanwhile, the simple yet primitively
powerful Frenzy hammers the
rolling, wedge-shaped Tazbot.

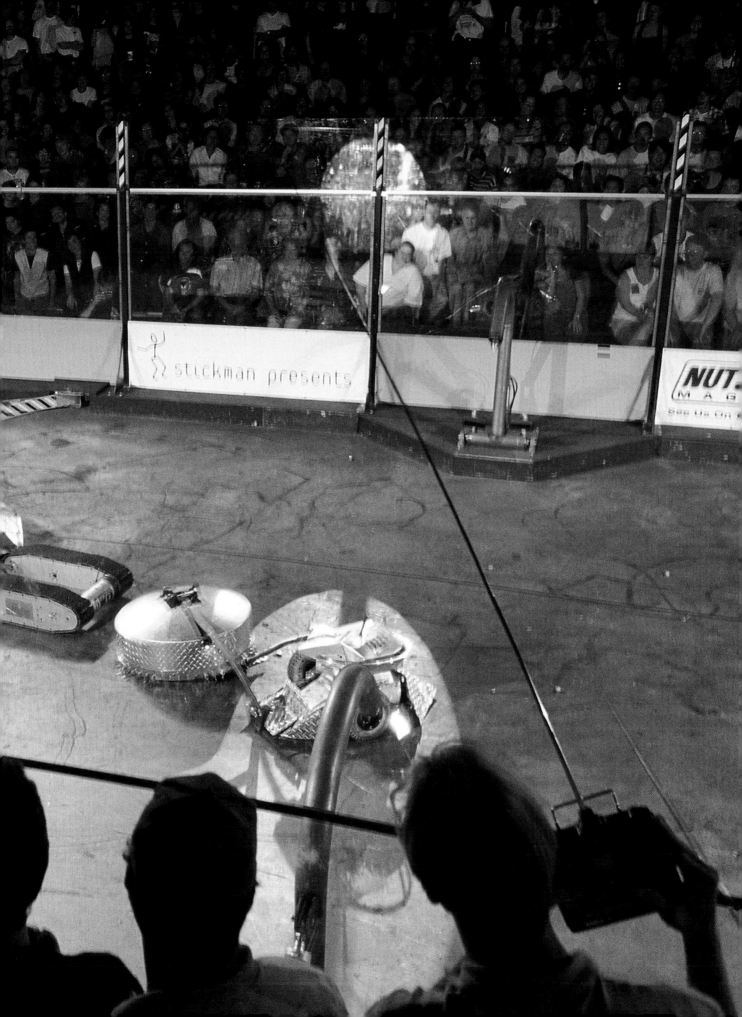

The most sophisticated machines don't necessarily triumph in the violent gladiatorial battles at San Francisco's Robot Wars, as shown when Tazbot (right, with turret), a simple, remote-controlled vehicle, forces a much more sophisticated, autonomously moving opponent to self-destruct. After the battle, robot owners quickly repair what they can in the adjacent pit area (above). Full of machines being groomed for combat and surgically rescued after it, the pit is a sort of electronic fighter's dressing room and hospital emergency room. Video monitors above the pit give contestants a view of the action. Returning from one-on-one matches, winners are greeted with high-fives and applause; losers are reminded that they have a chance to even the score in a final free-for-all, in which contestants in the same weight class battle to mutual destruction.

Turtle Road Killer, a sleek little robot that can turn on a speck of dust. Before the official robot weigh-in, its operator, Sam Spyer, 9, watches nervously to see what class Turtle Road Killer will end up in. "We're just going to make it," nervously promises Ted Taylor, a retired engineer who is Sam's grandfather. Team relief is apparent when Turtle comes in ounces under the limit. "It's the lightweight paint," deadpans Tom Petruccelli, the machinist/contractor hired to build the robot over a costly, work-intensive, three-month period. (Total cost to Sam's father: $11,000.)

Sam's battle partner, Taylor Tan, 9, has put Turtle Road Killer through its paces and thinks the robot can win. When it's their turn, the two boys guide their robot to the floor of the arena, then scramble up to the operators' platform on the side. They have agreed to take turns controlling Turtle's movement and the scooping paddle that flies up to flip unwary robot attackers.

Caught up in the excitement, they are unaware that many roboticists would not describe Turtle as a robot, because the two children directly control its actions. To purists, machines must move autonomously and react to their environment to earn the title of "robot." Sam, Taylor, and the rest of the robot builders at Robot Wars could respond, if they wished, that this definition would exclude the robots in robot car factories and the robots that NASA sends into space. They could also point out that if six thousand people are willing to pay $30 to cheer on a lot of robots, then what they're cheering on is probably a lot of robots.

When the two boys begin their five-minute match, the robot named Turtle whirls and flips its paddle while Hyena, its opponent, swings around in rapid defensive circles. The two machines thrust, hit, and spin, inflicting considerable damage on each other and sending chunks of metal spraying around the combat pit. When the two machines lock up in a steamy stranglehold, the strong-armed man guarding the entry door has to go in and pry them apart, something he ends up doing in contests throughout the two-day event. In the end, ultra-sophisticated Turtle is no match for Hyena, though it will go on to win its next round and show favorably in its final confrontation during the Melee.

Shortly thereafter, Lisa Winter's featherweight robot, Dough Boy, begins its match. A squat metal silver-colored box topped with a swiftly rotating lawn-mower blade, Dough Boy was a joint project between the precocious ten-year-old and her father. By herself, Lisa devised the robot's stealth component. She sewed small covers for Dough Boy's blades, she tells us, "so the other robot operators wouldn't know what they were fighting until the last

moment before the match." "It's a psychological edge," explains her mother. Her father says that the family pitted Dough Boy against a metal chair in a test run at their home in Monona, Wisconsin, and "the robot sent the chair flying about ten feet through the air."

After delighting the crowd by pelting them with fresh-baked crescent rolls that they could eat, Lisa wins her first match handily against a nearly immobile robot called Grinch. In fact, Dough Boy pummeled it savagely, much to the delight of the crowd. Lisa was unsurprised; she had scouted the competition earlier in the day and announced that Grinch was "beatable." After the battle, though, the blade on top of Dough Boy—the robot's chief weapon—stops working. She leaves the problem for her dad to solve while she goes off for a sleepover with a rival robot operator of about the same age.

As Lisa leaves the arena, the crowd is still roaring. In Robot Wars, humans have discovered a guilt-free way to indulge their perverse appetite for destruction.

Painted pink to give competitors a false sense of its harmlessness, Mouser Catbot 2000 has two deadly sawblades in its nose and tail and a hidden flipper on its back for overturning enemy robots. Built by Californians Fon Davis and April Mousley (above, left to right), the machine deftly trounced Vlad the Impaler, a larger machine with a hydraulic spike that shot from its snout.

Robo Specs

Name: Mouser Catbot 2000

Purpose: April thought it'd be fun.

Creative inspiration: Cute but vicious cat

Height: 20 cm

Length: 80 cm

Weight: 23 kg

Vision: None

Cost: $1,500

Information from: Fon Davis

Talking Heads

When Alvaro Villa flips though the photographs of the robots we've seen, he stops and stares in amazement at the robot cockroach from Case Western Reserve University (see page 102). "Is it for a movie?" he asks finally. "No," I tell him, "it's for research." Villa shakes his head in bewilderment. "It looks complicated," he says. By the time he comes upon the face robots in Tokyo (see page 72), he's ready for me. "Research?" he asks. His incredulity surprises me. The face robots don't look all that different from the faces of his own animatronic figures. The face robots are long-term research projects; you might find Villa's creations singing to you in the Tunnel of Love.

The term "audio-animatronic," invented and trademarked by the Walt Disney Corporation in the 1960s, refers to anthropomorphic or animal-like electromechanical figures with synchronized movement and sound. Villa's first memory of such figures is vivid—Christmas bears, built by the German company Himo, dancing in his native Colombia. Villa first encountered the business end of animatronic technology while working for Disney. After leaving the company, he founded AVG, Inc., in 1978. Since then, he has delivered thousands of animatronic figures to amusement parks, museums, exhibitions, trade shows, and shopping malls. His creatures, which act in movies and haunt amusement-park rides, range in appearance from cute to downright ugly.

In his Los Angeles-area headquarters, Villa shows off one of his favorite creatures—Crypt Keeper, the one-time host of television's *Tales from the Crypt*. Made to resemble a decomposing cadaver, it gives me the creeps when it isn't even powered up. The Crypt Keeper is but a small part of Villa's output. AVG has built animatronic dragons, insects, birds, and legendary heroes. Villa has even created animatronic flowers for his home country.

Alvaro Villa: While I was in Bogotá selling this idea of a flower show—that was a big project, 110 animated Colombian orchids—I was invited to a science show. They asked me to bring something robotic. We didn't have anything on hand so we made a coffee bean with three functions. It was so stupid, you wouldn't believe it. We spent a week and a half making this coffee bean—I put a Colombian hat on it—like Juan Valdéz [the trademark of Colombian coffee]. And there was some guy from France in the same science pavilion where I was who had this sophisticated robot that played soccer that they had advertised in

the paper. And when everyone came, nobody was watching that soccer robot—everybody was watching the coffee bean [laughs]. It made all the TV channels in Bogotá!

Faith: What did your coffee bean do?

It talked about the orchid show we were going to do, which was sponsored by the Colombian coffee growers. The guy who was in charge saw the public gathered around the coffee bean and says, "This is it!" We spent some money and rebuilt the coffee bean and made it the spokesman of the flower show.

Were you the kind of boy who took everything apart in your house?

Oh God! I was the terror of my family [laughs]. We had a box turtle as a pet for many years. It was pretty big. One day I put him in the front lawn of the house and put a speaker under it. I ran a wire to the house with an amplifier and when people were walking by, I'd say to them, "Hey, hey, I'm a turtle," and they thought the turtle was talking to them. I got in trouble for that.

If you weren't doing this, what do you think you'd be doing?

Maybe still working in the space program. I left the space program to go to Disney. When I went there, I knew I was home. I loved my time there. Nine years I worked with Disney.

What did you do for them?

When I first joined the company they were developing the first computer control system for animatronics. And in Orlando I did the electronic installation of some of the shows, like the *Haunted Mansion*. I did another one that was the first sophisticated animatronic computer-controlled show—*Bear Country Jamboree*.

How sophisticated can you get with the animatronics you're doing now?

The ghoulish host (left and above) for *Secrets of the Crypt Keeper's Haunted House*, a Saturday-morning television show for kids, is an animatronic—that is, lifelike electronic—robot. Built by AVG, of Chatsworth, California, the Crypt Keeper can show almost every human expression, although it must first be programmed to do so. Larger gestures of head and hand are created not by programming, but by electronically linking the robotic figure to an actor. As the actor moves and speaks the Crypt Keeper's dialogue, sensors on his body track the position and motion of his movements; a microphone records the accompanying dialogue. Technicians play back the recorded data, splicing in the appropriate creepy facial expressions.

Robo Specs

Name: The Crypt Keeper

Origin of name: Name of character played on TV show

Purpose: To host television program.

Height: 1.1 m

Weight: Approximately 27 kgs

Vision: None

Sensors: None

KLOC: Figure is proprietary

Project status: Complete

Information from: David Hall/AVG

Alvaro Villa translated his boyhood love of electronics into AVG, an animatronics company he founded in the Los Angeles area. Today he takes great pleasure in showing off animatronic figures like the Crypt Keeper (previous pages) or Little Man, the hip figure (above, with Villa) that "represents" the company at trade shows. Wearing a baseball cap and sneakers, it tirelessly delivers a humorous prerecorded spiel that is synchronized with a video on a screen behind it.

Wielding a specially adapted needle, Lyudamila Budnik (right) pushes eyebrow hairs into the head of a Greek princess at the Jacksonville, Fla., headquarters of the Sally Corporation, a manufacturer of animatronic figures. This princess will become a small part of Labyrinth of the Minotaur, an interactive indoor ride in a new Spanish theme park, Terra Mítica (Mythic Earth).

The majority of people don't want to pay for sophistication because the majority of people don't need it. When you do a dark ride like we just did for Taiwan—

That's your latest project?
Yes. It's a shipment of dragons and human figures for a Taiwanese amusement park. When someone goes through a dark ride [a dimly-lit indoors amusement-park ride, usually intended to scare the customer], they don't have a lot of time to focus on one figure. You want them to go through there and get the idea that these guys are alive and are doing something that relates to whatever the story is. So we don't need so much sophistication. If it's one figure and you're going to be looking at that one figure up close for a long time—like the Crypt Keeper on television— you'll be picking up on all the details, so we give it more functions. But what the market is willing to pay is a different thing altogether. That's why we do things the way we do them. Because if we cannot sell it, it doesn't do us any good to do the research.

But you can take advantage of other research that's done.
Yes, yes. For example, I've been playing with this technology called "muscle wire" for about three years to find a practical way to use it. I have a little sample over here and it's been running for about two years—I bought a lot of samples. But it has a lot of restrictions: it runs a lot of current and the recovery time is very slow. Everything that comes out new— we buy it and we play with it—especially if we think we can apply it.

Do you build sets and props as well as the animatronic figures?
Sometimes we do everything.

This is a room full of boxes of machined parts. What are we looking at?
For example, these are head plates for human heads. They're adjustable, so you can make a child or a grownup. We also have finger parts that should be right here in this box, but there aren't any left. We just used a lot of them for another project.

Do any of your animatronic figures walk?
We don't make the walking robots because it's very expensive and nobody needs it. We have to make things that people want to use.

There are a lot of roboticists out there working on walking robots. What do you think about humanoid robots?
When we did work for Singer [the sewing-machine company], they thought we were going to make a human-type robot that would take a piece of material and start sewing. We did the opposite: we built a machine to do this because we knew how complicated we humans are. We made a machine that has an alignment system with a flotation table to float the clothing and align it before picking the clothing up with an end effector [hand-like gripper]. The human-type robot doesn't, to me, have a practical purpose.

Do you keep abreast of the whole robotics field?
Not as much as I'd like to. I look on the Internet but I don't do as much as I'd like to, because I also have to run the company.

Do you see yourself as a researcher?
Oh, absolutely, yes.

From the time of the turtle?
Yes, from the time of the turtle [laughs]. I have been forced by life, in whatever research I do, to make money from it. We develop all our actuators because the ones we can buy are not efficient to place inside a humanoid figure.

Economic necessity pushes you to push the envelope.
Exactly. And the same with the valves. The pneumatic valves available on the market today are hydraulic valves converted to pneumatics. They are big and heavy and inefficient, so we make our own.

Oblivious of the seeming threat from *Tyrannosaurus rex*, staff artist Lance Jacobson sculpts a batlike creature in one of the Sally company's large, eclectic workspaces. The bat is destined for an interactive amusement park ride in Terra Mítica, a new Spanish theme park. In the ride, patrons will sit in automated, chariotlike vehicles that rattle through a mythical Greek kingdom. Armed with low-intensity, laser-firing crossbow pistols, they will take potshots at dozens of animatronic evil doers and monsters—among them, the bat-thing Jacobson is working on today. "This is just a you-shoot-it-and-the-light-goes-out type thing," he says. "These will be just hanging on the cave walls all over the ride." Surrounding the artist are full-size foam models of human beings, parts of which will be animated to make them appear lifelike. On the desk in front of Jacobson, a plastic toucan perches on a metal stand, its robotic innards clearly visible through its transparent skin.

Electronic Kicks

Robo Specs

Team name: J-Star 99

Origin of name: Acronym for Joint Strategic Technology Alliance for RoboCup 1999

Purpose: RoboCup small-size league competition

Height: 15 cm

Length: 18 cm

Vision: Camera mounted over field

Sensors: Infrared

Frame composition: Aluminum body, carbon-fiber cover

Batteries: Custom

Project status: Ongoing

Information from: Hiroaki Kitano

Team name: CMUnited-98 (see page 214)

Origin of name: Carnegie Mellon and the fact that it's a team, therefore united. We had a winning team in 1997, therefore CMUnited-98.

Purpose: To play at RoboCup— a testbed for research in multi-agent planning, execution, and learning. They were built to be a *team* of robots that could achieve concrete goals in an adversarial environment, like defending, passing, and scoring goals in robotic soccer.

Project status: After a very interesting demonstration at the Smithsonian in December 1999, we completely retired this set of robots. We are currently building new hardware for a new team of robots.

Information from: Manuela Veloso

Team name: GMD RoboCup (see page 215)

Origin of name: GMD is the abbreviation of the research facility name.

Purpose: To play at RoboCup.

Height: 45 cm

Vision: 360-degree freely turnable camera with 3 color channels

Sensors: Odometer, bumpers, short-range infrared, long-range infrared, gyroscope

Frame composition: Aluminum

Batteries: 48 V 1.2 Ah

External power: None

KLOC: 35.4 (microcontroller), 34.2 (simulator), 15.2 (TraceGui)

Cost: $12,500

Project status: Ongoing

Information from: Andrea Seigberg

Bob Walter and his son Oscar are joining friends tonight at the World Cup Soccer championship at Microsoft's Disney Memorial Stadium in Beijing. Oscar is at college in New York City; Bob is at home in Berlin. The two meet at Bob's house—virtually, of course—exchange greetings, and then switch over to the game site via the FOXyNBC virtual-presence link. Their friends from TWACK (TimeWarnerAOLCBSKelloggs) are waiting. "You're just in time," they say. "It's about to begin." Music fills the stadium, or so it seems, and the teams are announced. Before a sellout virtual crowd, two teams take the field: eleven lean, athletic-looking human players and eleven sleek, athletic-looking robots. The RoboCup Championship of 2090 has begun.

Actually, says Hiroaki Kitano, human-versus-robot soccer matches may occur before 2090. As co-founder and president of the RoboCup Federation, the rapidly growing world association of robot soccer teams, Kitano has been working toward this goal for years. "Our final goal is that by the year 2050 a fully autonomous robot team will play the human World Cup champion team," he tells me. "And all the normal regulations will apply."

"Isn't that dangerous?" I ask. We're talking in the office he uses for his day job as head of Kitano Symbiotic Systems, a private, nonprofit robotics R&D facility. "With the robots we've seen, you wouldn't want to have one smash into you."

"The problem will be to convince them to play against the robot," he admits. "That means we'll have to make really sure that the robot is very safe and consistent."

Right now, the notion of robot-human competition is the stuff of sports science fiction. In today's RoboCup, the miniature mechanical teammates strain their capacities just being able to recognize other teammates, let alone pop the golf-ball-sized ball into the goal. Fielded by university faculty and student researchers from around the globe, the five-robot teams are intended to function as units, blocking, passing, and setting up goals. In a traditional academic environment, this kind of research would be called "multi-agent robotics"—groups of robots that work together on tasks. And, in fact, Kitano and other RoboCup organizers expect that their sports work will eventually be a force pushing the development of such multi-agent endeavors as landmine clearing and search-and-rescue.

Faith: Why did you pick soccer?
Hiroaki Kitano: What we are trying to do is use football—"soccer" in the United States—as the way to drive a technology. Soccer is a kind of universal language. In some countries it's a kind of religion. The passion for football is one of the driving forces for the research—everyone gets excited. One surprising thing I saw in the 1998 RoboCup competition in Paris was that teams from Italy came over with about 30 or 40 students with a national flag. That was the first time I saw people cheering and waving a country's flag at an academic conference [laughs].
You have to be careful that nobody brings some robot hooligans.
Germany has five or six middle-sized teams and a dozen or so simulator teams. Obviously, the Germans are very serious about this. And I was surprised in the 1998 RoboCup Pacific Rim Championship in Singapore that many universities and polytechnics had RoboCup teams.
Now, the competition mainly features small robots. When do you expect to put full-sized humanoid robots on the soccer field?
In 2002 we're going to have a humanoid robot-biped class in the competition, although we will introduce it earlier. We'll probably have small-, medium-, and human-sized humanoid leagues. I'm not sure how many teams we'll have. That will be very expensive and challenging, to create a biped robot that can kick the ball.
It doesn't seem just too difficult to even think about now?
It's important just to start. No matter how problematic it is at the beginning, you have to just push the technology. What is interesting about RoboCup is that if you just write a paper and present a result in a conference, you can test your robots a hundred times, achieve the result one time, and present that sequence in the video and show that you did it. In RoboCup you have to carry the robots and the technology overseas and do it live in front of many people—in front of CNN, newspapers, magazines, and other researchers. You can't plan ahead of time and do it once. You must have a very reliable technology. Eighty or ninety percent of the time, you have to have your robots up and running.
That's not good enough for robotic surgery—
Yeah, yeah [laughs]. There it is a life-and-death matter.
Which team impressed you the most early on?
Carnegie Mellon [University]. The group could actually pass around the ball. They could move the robot around to do a back pass. That was really amazing, to see that so soon!
It was more advanced than you thought would be possible at the beginning?
We thought that in a few years someone should be able to have a system like that, but we didn't expect it to happen in 1998. It takes a lot of preplanning but they also have to have a lot of real-time maneuvering.
And there was no human involved?
They were free autonomous robots.

北 1

In a spanking new, richly appointed research center above a busy shopping street in Tokyo's stylish Harajuku district, Hiroaki Kitano (left) shows off his robot soccer team. In addition to Kitano's humanoid-robot work at Kitano Symbiotic Systems Project, a five-year, government-funded ERATO project (see page 52), Kitano is the founder and chair of Robot World Cup Soccer (RoboCup), an annual soccer competition for robots. There are four classes of contestants: small, medium, simulated, and dog (using Sony's programmable robot dogs). Kitano's small-class RoboCup team (above) consists of five autonomous robots, which kick a golf ball around a field about the size of a ping-pong table. An overhead video camera feeds information about the location of the players to remote computers, which use the data to control the robots' offensive and defensive moves.

Reviewing the results of her work, Carnegie Mellon computer scientist Manuela Veloso (above, kneeling) watches the university soccer-robot team chase after the ball on a field on the floor of her lab. Every year, the Carnegie Mellon squad plays against other soccer-robot teams from around the world in an international competition known as RoboCup. Veloso's team, CMUnited, is highly regarded; it was world champion in the small-robot class two years in a row. Flanked by research engineer Sorin Achim, postdoctoral fellow Peter Stone, and graduate research assistant Michael Bowling (right to left), Veloso is running through the current year's strategy a month before the world championships in Stockholm. Since the last competition, coach Veloso has spent countless hours working out software bugs and hardware glitches while trying to outguess the innovations of competitors from a dozen other countries. CMU's AIBO team members are Scott Lenser, Elly Winner, and James Bruce.

The team Kitano so admires is led by Manuela Veloso, a computer scientist at Carnegie Mellon University in Pittsburgh. When we drop by, Veloso is putting her robots through their paces with graduate students. As the small robots, each the size of a human hand, shuttle around the makeshift field, the high-energy Veloso barks orders at them—a futile exercise, given that the machines have no capability to hear her. "No, no, no, no, no! Not that way—where are you going? Where is he going?" she asks no one in particular. Veloso's competitive fire is apparent, even when her robots are practicing with no opposing team.

Peter Stone, team member: With one robot working alone, it's extremely difficult to score.

Manuela Veloso: It's very hard unless you do some passes. Our vision is actually a prediction.

Faith: Your vision systems project the ball's trajectory?

Veloso: The goalie can use the predicted trajectory to move to a place where the ball is going to enter. Therefore, it's hard for these guys to score by themselves. It's really hard. I think it won't score by itself [referring to a robot moving forward toward the ball]. There we go—it kicks and the goalie moves exactly to the right point to stop it. And here is a dangerous situation! It tries to go around!

Michael Bolling, team member: The goalie or the

wall—it's not given any other option for a place to shoot. Whether or not he is going to shoot, the goalie is going to get it anyway.

Veloso: We'll try to make a pass to him. Come on! Open up!

Stone: Actually you could make the guy score. Something set up in front, I think you can—

Veloso: No, no, not really.

You sound like you're coaching human players.

Bolling: We yell at them.

Veloso: We yell, "Go! What are you doing?"

Stone: We once called a robot lazy, I think.

Bolling: We always have [Robot] Nine play. Nine works better, we claim.

But they're all exactly the same, except for the goalie.

Veloso: Yes, but we still say this. And we kind of fell in love with [Robot] Seven. We say, "Oh, Seven needs to be in the midfield." Nine was the defender. *Why?*

Veloso: He was a very good defender.

How does your RoboCup research differ from other robotics, do you think?

Veloso: In most robotics people worry about obstacle-avoidance for stable obstacles. Robots have to go around the room, but the tables don't move. It's no big deal to go around objects that are stopped. You learn they are there and then you never go

wrong. These guys here are moving and everybody else is moving, so you are constantly having to do this path planning.

Sony also gave several research groups, including yours, sets of the AIBO [the company's robot pet] for a four-legged robot-soccer league.

Veloso: For this one the challenge is different because they have everything onboard. They have a camera, they see, and they move. Their angle of vision is quite wide and they can see well. The challenge here is to know where you are on the field. That's why they have colored landmarks. So we have to develop an algorithm to figure out "If I see this marker and that marker, I am probably somewhere in this circle, and therefore I should be in front of the goal." Those guys have a view of the whole field. But they cannot communicate. They don't do any teamwork. They are autonomous. When you see them playing, it's awful, because they all go for the ball.

What are some of the difficulties the robot faces that a human wouldn't?

Veloso: A robot searches for the ball, finds it, circles around until it is aligned with the goal, and then tries to kick it into the goal. But as soon as it sees the goal, it doesn't see the ball anymore—and when it sees the ball, it doesn't see the goal any longer. So we keep having to build memory, so that it can

remember things it is no longer looking at. Humans have an amazing ability: "I'm not looking at my house, but I know where it is. I'm not looking at my car, but I know where it is." Unless we build the memory, these guys forget.

Does it have a memory per se?

Veloso: Well, it's cumulative. They have a belief of where they think they should be. They have to keep remembering where they were. If they have the ball in front of them, they look up and see the ball. If someone comes and takes the ball, they will wonder, "Well, the ball was right here. What happened?" And then they start again. They look for the ball—where's the goal? The cycle starts again. So if you see the games, they are very slow.

Part of the memory challenge is not just remembering where things are—you can always remember where things were. The question is deciding when to forget. Because a ball moves, whereas a house stays still. If you can remember where the house is, that's it. But if you don't see it, the ball sometimes might be where it was and sometimes it might not be where it was. So people are very good at saying, "Oh, well, it's been so long since I've seen it that it's very unlikely that it will be there, unless there is nobody else around kicking it and I haven't been kicking it, so it must be where it was." To encode this into robots is something we are constantly trying to struggle with.

Readying for the RoboCup championship in Sweden, Jörg Wilberg (above, rear left) and his research team at the German National Research Center (GMD) outside Bonn review the prospects of their five-machine robot-soccer squad. The GMD team plays in the medium-sized division, which uses a real soccer ball on a field about a third as big as a basketball court. Each robot monitors the position of the ball with a video camera; special software lets the machine track its round shape. Kneeling on the floor, researcher Peter Schöll tests the software by observing the image of the ball in the monitor.

Tech Talk

From RoboCup Web page (www.robocup.org)

The RoboCup Federation will start a Humanoid League beginning with RoboCup 2002 in Japan. Initially, each team will consist of up to three robots, but the number will increase to five, seven, and then to eleven with maturity of the technology. It is assumed that these humanoids are fully autonomous.

A Head of His Time

Steve Jacobsen is a roboticist's roboticist. Although he is not the most well-known researcher in the field, Jacobsen is certainly one of the most prolific. A professor of mechanical engineering, bioengineering, computer science, and surgery at the University of Utah, Jacobsen is an entrepreneurial researcher who, along with a bright team of researchers, operates out of the Salt Lake City headquarters of SARCOS, the robotics company he founded in 1983.

Hopscotching from thoughts about the future to discussions of the impact of dyslexia on its victims'

personalities (Jacobsen is dyslexic), he describes his ideas virtually nonstop as he takes us around the SARCOS headquarters. Jacobsen, a polymath, is well versed in physics, biology, engineering, and any number of other subjects. If that isn't enough, Jacobsen is a wellspring of anecdotes: about everything from making robots for Disney, to creating programmed fountains for Las Vegas, and designing robotic limbs for amputees—the Utah Artificial Arm (see page 178) is worn by thousands of people.

Although the building is jammed with whimsical animatronic figures SARCOS has built for theme parks and stage shows, Jacobsen's greatest contribu-

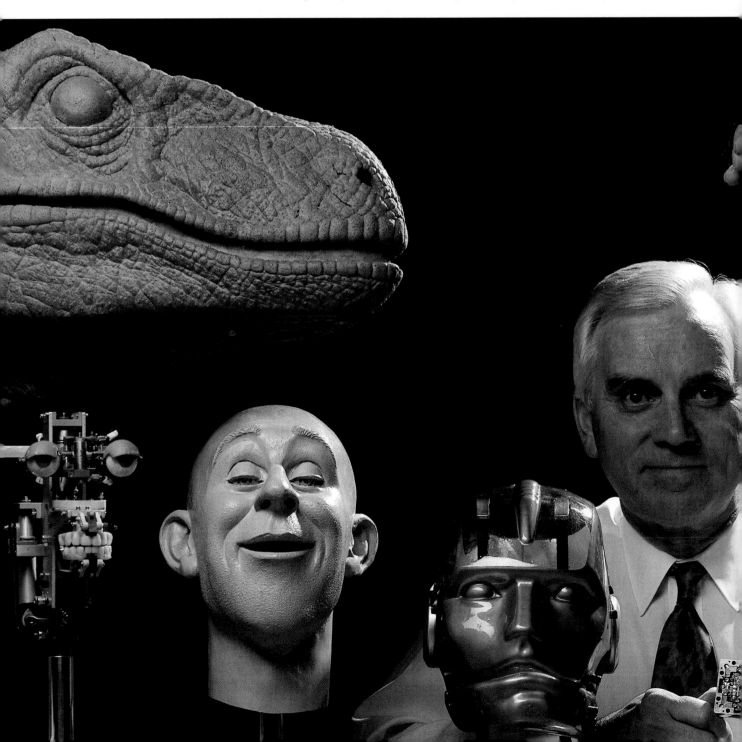

tion may be his medically oriented devices like artificial arms, artificial kidneys, and patient monitors. Downstairs in a dust-free, glassed-in "clean room" is the SARCOS chip-fabrication facility, a place for complex industrial production. In the next room, personable animatronic parrots that sing "Feliz Navidad" are being created by SARCOS for a local restauranteur. After Jacobsen gives us the tour, I ask him about the difference between working in academia and working in a commercial company.

Steve Jacobsen: Many times in academia, if something wiggles and does stuff, people think it's really

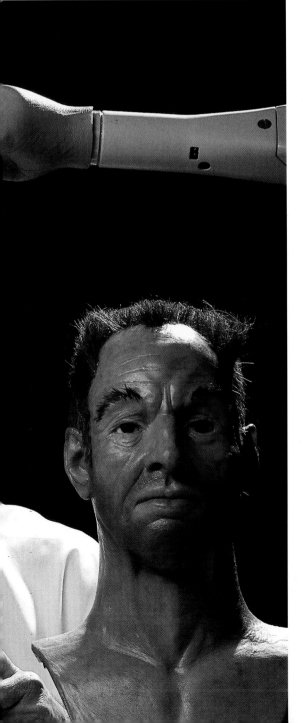

wonderful, but it really isn't a proof of anything.
Faith: You can't sell something that only looks good—it has to work.
Making something industrially pulls you out and makes you deal with reality. Doing this stuff is like walking carefully in a mine field. One wrong step—it blows your wallet off.
Faith: Can you imagine a day when robots will be advanced enough to build themselves, autonomously?
Why would you ever self-replicate? I mean, we'll build a lot of robots that are machines, but the idea of having the complexity required to self-assemble and replicate? I think the first self-replicating robots will be done by geneticists, not by mechanical engineers. Biology is different. There are rules in biology—we just don't know them. You have to do procreation, because you have to mix things genetically, so replication is part of the package from the beginning. It's scary, isn't it? I mean, you look at stuff that's going on in biology and genetics—
At the same time, you're obviously influenced by biology.
You have to respect the power of nature. Nature has created an enormous number of solutions to problems, and we should take what we can from it. My idea is turning machines into what we call "organismic" machines. Organisms like you and me have sixteen different types of skin sensors, all sorts of muscle sensors. You have sensors everywhere and actuators everywhere and they're all working together. And you can tell by the law of conservation of difficulty—
What's that?
If there's a problem that hasn't been solved, then there's usually something about it that's tough. It's one of the things that designers say.
In this case, you're saying that biological systems are incredibly complicated to understand, so it won't be easy to duplicate them.
You are not going to get machines to be amazing until you do the same kinds of things that nature does to make them work.
Our bodies are like incredibly complex machines. Maintenance is critical for any organism, both biological and mechanical, right?
There's a big thing in the military called condition-based maintenance. By maintaining something by a book or by pulling Gs in a fighter plane by a book, it tells you what the limits are. But if you put sensors in the tail that say when you're pulling six-Gs "I'm OK, I'm OK," you can pull seven. Another thing they want to do is reduce manpower on ships. Almost everywhere, they're trying to extend capital assets—life extension of capital assets. So, how do I make a plane last twice as long? Or how do you keep from breaking it, or how to know when to maintain it, or how to operate it closer to the limits? So that if someone is after me, and I can pull more Gs, it all comes down to knowing about the machine. And once you have done

Peter's Notes

From the picture on this page, one could dismiss a lot of Steve Jacobsen's work as "automata"—mechanical figures that simulate the motion of living things. The oldest form of robotic technology, automata were designed by Leonardo da Vinci 500 years ago. Leonardo's device was supposed to mime the complex motion of a bird's wing. Many of Jacobsen's robots do similar things, only in more complex and sophisticated ways. The true measure of his genius, however, is suggested by the 124 patents he has received during the last 25 years—everything from an articulated prosthetic wrist and an implantable drug-delivery system to a system for continuous fabrication of micro-structures and thin-film semiconductor devices. The headquarters of his company in Salt Lake City is as eclectic as its founder: a room full of choreographed robotic parrots is next to micromechanical testing and fabrication labs. Across town is another facility where his company builds everything from humanoids to one-off products like the robotic T-Rex for Universal Studios and the dancing fountains at the Bellagio resort in Las Vegas. If Jacobsen and da Vinci had been contemporaries, they would have been pen pals.

Flanked by the animatronic robots created in his workshops, Steve Jacobsen (left), an engineering professor at the University of Utah, may be the world's most entrepreneurial roboticist—he's spun off four companies from his research and discoveries. Using the sophisticated electronics Jacobsen and the SARCOS team has invented—an example is the 6-axis force/torque sensor (left, in his hand)—his companies have branched into microcomputers, digital and analog controllers, sensor networks, and programmable drug-delivery devices. But perhaps the most important product he makes is the Utah Artificial Arm (above Jacobsen's head), a high-tech prosthetic hand used by thousands of amputees around the world. Inside this myoelectric (muscle-electric) device are several small stainless-steel electrodes that contact the user's upper arm. The electrodes read the electrical impulses in the arm muscles, amplify them, and transmit them to the circuitry in the mechanical hand, which moves in response. After training, amputees can use their Utah Artificial Arms to flex, rotate, and grip with considerable dexterity.

this, almost every machine becomes a robot in the classical sense. It has sensors, actuators, brains, and it responds to situations. But one of the problems in robotics is that we've clung to the notion that it's got to look like a person, and if it fails, they say, "See—robots failed," and yet they are all around you.

I see what you mean. We could be surrounded by robots and hardly even know it. They could even be inside of us—you're building these tiny, experimental drug-delivery robots that will be implanted in the body.
We're building new micro pumps now that are way small. This is the smallest pump in the world that's practical [picks up device from his desktop that is smaller than a matchbox]. It's very accurate—it's got a microprocessor-drive motor. People say, "That's not a robot." But it is—it has actuators and sensors, it has brains, and it's got a job. We're looking at using these systems for diabetes. The pump has to sit there and look at what has happened to a diabetic's glucose, look at the time of day, perhaps, maybe get something from the patient about what they are going to eat. Then it makes a decision whether to give insulin. And if it does a good job, you help the diabetic. It's all about keeping homeostasis. Because when diabetics get big swings of glucose they damage cell membranes, and that's what causes the problems with the eyes and appendages that plague diabetics.

What else are these kinds of little robots good for?
We did a personal-status monitor project, which was for monitoring soldiers in the field. It has gone up and down but is still alive and well with several projects. One with the Navy is called Man Overboard. Another is called First Watch. We also have a joint project with Draper Labs at MIT, monitoring sailors. Say you are in battle conditions and you close all the hatches—like when an aircraft carrier goes on alert, you have a certain amount of time to get out and then the doors close and they do not open again. If you have damage, you want to know who is where and if certain areas can be reopened to get to the injured.

So all of the sailors have personal monitors worn on their bodies that provide this information?
These, again, are turning a ship into an organism—turning a ship into a robot that has self-awareness, where people are just little cells running around in there.

You mentioned using this type of system to monitor soldiers in the battlefield? Is it for helping medics find injured soldiers, or something like that?
The medic has a small monitor with him, giving him two kinds of data. First it will point an arrow to where the guy is hurt and how far he is from the medics. Physical status is analyzed in stages. First whether the soldier is alive or dead, then severity of injury, and then comparative severity of injury, triage—it has to be a double triage now. When you triage in a hospital, the only person in danger is the patient. If you are going out to work on guys in a

battlefield the medics are in danger, too. How bad off is this guy and how bad is the risk to a medic? If the medic can't go but he knows there's bleeding, or somebody's going into shock, he could push a button that could instruct the monitor to insert the needle and push another button and give the drug. We made these devices. They use micro explosives and little tiny incendiaries. A little chip system tells them when to go off.

Like tiny bombs to give people drugs?
They don't really explode—pressure is built and it ruptures the front. With more sophisticated personal status monitors, if all the troops carry external sensors they become an organism—a wide-aperture

array. You could tell the men to deploy in a line and put out their chemical sensors to triangulate where a chemical threat is coming from. Stick vibration sensors in the ground to find out where the tanks are coming from. The possibilities are huge.

Are these robot monitors only intended for soldiers?
That one there [picks up another small device] is for a cow, to give it antiparasitic therapy for the two months before you slaughter it. Every two weeks, it puts out a chunk of a drug called Ivermectin, and by the time twelve injections have occurred the animal is parasite-free. When they slaughter the animal, they know where these drug-delivery devices are, and they're not metal so they won't wreck their machinery.

It's amazing to think of these little robotic devices implanted in people everywhere. On the other hand, it's true that robots in general are more and more common, so maybe one shouldn't be surprised to think of small ones everywhere.
You maybe ought to be the first one to point out a few of these little things and say, "You know, they are all around us." It's like the *Body Snatchers* [movie] when the guy says, "They're all around us!" Not only are anthropomorphic robots doing better, but things of everyday usage are becoming organisms, machine organisms. Some are so big it's hard to think of them as robots, like aircraft carriers, but some, like these tiny drug pumps, are robots too.

In a photo-illustration, SARCOS, (above, cheating) an animatronic robot built by the SARCOS Research Corp., a Salt Lake City robotics company, appears to peer at the seven-card-stud hand of Scott Reynolds (above, at right), one of the engineers responsible for creating him (with technicians Doren Prue, center, and Charles Ledger). Like many entertainment robots, this one does not sense its environment or react to it. Instead, it either follows the script of a computer program or is remotely controlled by an operator wearing a sensor suit.

Tour de Force

Robo Specs

Name: Sweet Lips

Origin of name: A fish in the movie *A Fish Called Wanda*

Purpose: 1) To excite and inspire the public about science and technology and, specifically, robotics. 2) To educate and entertain museum visitors and attract them to exhibits in the museum that they might not visit. 3) To create a demonstration platform to prove that robotics technologies have matured sufficiently to survive, over the long term, all alone in a public space. These robots plug themselves in when they need juice, and page us and E-mail us if they need help. 4) To create a forum for us to conduct leading-edge research in human-robot interaction.

Height: 1.9 m

Weight: 115-135 kg

Vision: Landmark recognition, two single-board CCD chips, color, auto-shutter

Sensors: Touch-sensitive sensors around entire torso, 48 sonar sensors, 48 active infrared sensors, 8 real-time encoders

Frame composition: Anodized aluminum

Batteries: 8 sealed lead-acid, deep-cycle batteries, a total of 240 amp-hours of energy capacity (!)

External power: Optional 120 V AC

KLOC: 30.4 C/C++

Cost: $45,000 (components only)

Project status: Definitely ongoing.

Information from: Illah Nourbakhsh

Peter's Notes

We spent a morning in the Carnegie Natural History Museum watching Sweet Lips give tours of the animal dioramas. Kids love interacting with the robot, which runs flawlessly. But at times it gets lost, runs into walls, or columns—mostly due to mistakes in the computer code, and occasionally due to changing light conditions that confuse the robot's vision system. Sweet Lips' older brother, Chips, works downstairs in the dinosaur hall; its younger brother, Joe, gives tours at the nearby Heinz History Center. Joe is different: it has arms, and an attitude. Illah Nourbakhsh has irreverently programmed it to be slightly cynical. "It's a robot," he explains. "It can get away with that, whereas people might get upset if a human tour guide behaved that way."

A little boy stares directly into Sweet Lips's face as the robotic museum tour guide describes the fishing techniques of bears. Although a video screen in the machine's chest shows an action-packed bear video, the boy is transfixed by the robot itself. The reaction is not uncommon—Sweet Lips has been fascinating tourists for more than a year at the Carnegie Museum of Natural History in Pittsburgh. Coming into contact every day with museum visitors and staff, it is one of the few autonomous robots that functions in the real world every day.

I spoke with the machine's principal designer, Illah R. Nourbakhsh, chief scientist at Mobot, Inc. and professor of engineering at Carnegie Mellon University.

Faith: This is one of the few robots actually working among people on a daily basis. Judging by the difficulty other roboticists have had working toward this goal, you've achieved something monumental.

Illah Nourbakhsh: If making a robot actually move about with people and interact with them is like climbing Everest, then having a robot do this all by itself, day in and day out, is like traveling to Mars! The problem is, a robot that is 90 percent reliable can be built with some hard work, but a robot that is 98 percent reliable takes much more effort. A robot that is close to 100 percent reliable takes not just intense progress, but a few small miracles, too. There are some robots in controlled environments like factory floors and hospital corridors that deliver packages reliably. But doing this in the middle of a public space like a museum is a whole new problem, and that is what makes Sweet Lips so special. Sweet Lips proves that we finally have the technology to make real, autonomous robots that interact with people all day. You can expect to gradually see more such robots over the next few years, popping up in shopping malls, sidewalks, science centers, and airports.

How smart is Sweet Lips?

This robot takes care of itself: it plugs itself in when its batteries need to be charged, and it keeps track of its own tour schedule, unplugging and providing tours whenever the museum is open. Sweet Lips can actually determine when it needs help. If something goes wrong, it will try to deal with the problem itself and if that doesn't work it will page and E-mail us to ask for help. It is a big step forward when an autonomous robot can decide to ask for help—that introduces a level of self-sufficiency and awareness that is new.

Can Sweet Lips get lost?

It has gotten lost a couple of times. The most com-

mon culprit: burned out lights in the museum cause Sweet Lips to not see quite what it expects with its video camera. In one case its older brother Chips became totally lost and E-mailed us for help because the museum lights were accidentally turned off before it had plugged itself into the wall.

Our robots do not carry flashlights, and they use vision to see the world; so, when the lights are out the robots are blind.

What are your plans for the robots?

Unlike projects that aim to send robots to the Moon or to Mars, our goal is to create self-sufficient robots that are socially active members of society right here on Earth, in our cities and throughout our buildings.

[*A museum guard tells me what it is like to work with an autonomous robot, though she doesn't want to give me her name.*]

Faith: You are one of very few people who work full-time around an autonomous robot, so I'm interested in what you think about it.

Guard: Oh, I think robots are going to take over.

Do you really? Just by being around this one?

This one here is good. She comes out and if you are in her way she'll tell you, "You are standing in my way, you are standing in my way."

What happens if you don't move?

She'll keep telling you and you'll get so tired of hearing her say it that you'll move.

It stands in its battery-charger overnight and comes out in the morning?

Yes. You've got to watch her come out, it's amazing. I always watch her come out of that thing. Her eyes flash and she just moves out. And she snaps herself right back in at 4:30.

So, generally, you like the robots.

This is a nice one. I kind of like her.

Why?

Oh, I just like her. You should see the kids run up to her. They follow her around, they talk to her, they press her buttons.

Sweet Lips is what I'm told she's called.

I didn't know that. Well, she doesn't have any lips.

Sweet Lips the robot guide (above) takes visitors through the Hall of North American Wildlife, near the Dinosaur Hall in the Carnegie Museum of Natural History in Pittsburgh. Carnegie Mellon University robotics professor Illah R. Nourbakhsh's creation draws children like a pied piper by speaking and playing informational videos on its screen. It navigates autonomously, using a locator system that detects colored squares mounted high on the wall. A color camera and scores of sonar, infrared, and touch sensors prevent Sweet Lips from crashing into museum displays or museum visitors.

Robo *sapiens*

Roaming the sands like a glowing desert scarab, six-inch-long
Unibug 1.0 strides across the
wasteland of the Great Sand
Dunes National Monument in
the Colorado Rockies.

PREVIOUS PAGE

Passersby amble past the friendly
robot giant at the doorway of
FAO Schwarz in Orlando, Florida.

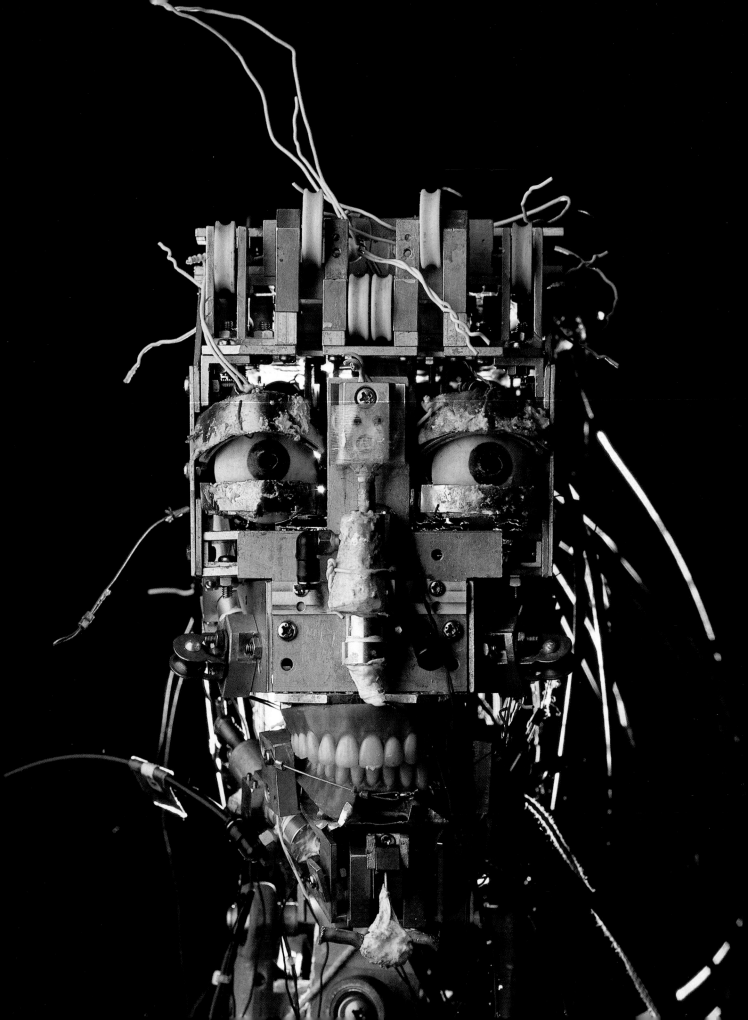

Even someone who believes that in the future most humans will become the slaves of all-powerful machines has to have a laugh sometimes. Why not have it with toy machines? Taking a moment off from his work at the cybernetics department at the University of Reading, Kevin Warwick (on left), author of *March of the Machines: Why the New Race of Robots Will Rule the World*, plays with Lego Mindstorm robots that his students have programmed to box with each other. The toys are wildly popular with engineers and computer scientists because they can be programmed to perform an amazing variety of tasks. In this game, sensors on the toys determine which machine has been hit the most. In his more serious work, Warwick is now trying to record his neural signals on a computer and replay them into his nervous system. In this way, he hopes, he will be able to drink a glass of wine on one day and the next day recapture the flavor with Proustian fidelity.

A Dog's Life

Robo Specs

Name: AIBO

Origin of name: The first two letters of the word AIBO stand for Artificial Intelligence. AIBO is also a robot with eyes, so [it is] an eye-bo(t). Finally, AIBO is also named after the Japanese word for 'pal.'

Height: 266 mm

Length: 275 mm

Weight: 1.6 kg

Vision: Digital camera

Sensors: Contact, infrared, acceleration, angular velocity, microphone

Batteries: Lithium ion

External power: AC recharger

Cost: $2,500

Project status: Just the beginning.

Information from: Merran S. Wrigley

Name: Tama (see page 227)

Origin of name: Japanese cat name

Purpose: Healing human mind and having fun.

Creative inspiration: Real cat

Height: 25 cm

Length: 35 cm

Weight: 1.5 kg

Vision: None

Sensors: Tactile, auditory, posture

Composition: Artificial fur, plastic, aluminum

Batteries: Lithium

Project status: Ongoing

Information from: Takanori Shibata

Peter's Notes

I lived with a Japanese family in Tokyo for a week while photographing them for my first book, *Material World*. The first time I used their facilities, I was warmly greeted by an electrically heated toilet seat. Sounds crazy, but it actually made sense in their cold house where only one room had heat. I watched way too many hours of TV with them and was surprised to see dozens of toilet paper ads with little cartoon fannies doing what little fannies do. On the next trip I stayed in a big hotel with a robotic toilet seat—a weird little arm with a variable temperature, force, and pattern water sprayer could be guided around from a control pad attached to the toilet. Whoa! But the grand finale was a blow dry. So when I first saw this little AIBO robot dog running around the apartment pretending to pee by raising one leg, I was amused, but not surprised.

The Nozue family fish tank runs the length of the living room in their small apartment in Yokahama, Japan. In it lives their sixty-centimeter-long pet fish. After four years, Mitsuhiko Nozue dryly observes, "it still doesn't know me."

Pets of the four-legged flesh and blood variety are not allowed in the Nozue family's apartment complex in Yokahama, Japan. So when Sony announced a new robot that looked and behaved like a small dog, Mitsuhiko Nozue decided that he, his wife Yahini, and their son Masahiko, 7, should have one.

There were roadblocks to ownership, however. Sony was selling only a limited number of the robot pets in Japan, and these could be purchased only by the winners of a special national lottery. Even then, winning the lottery meant winning the right to fork over $2,500 dollars for the mechanical pet, which Sony had named AIBO.

To their surprise, the family won and purchased the robot. They named it Narubo. When we visit them, I ask if they both wanted it. Mitsuhiko turns to Yahini and cryptically says, "Your opinion." Their laughter fills the room; obviously, they have discussed this subject before. As we continue talking, AIBO trills away with internal electronic sounds; Mitsuhiko watches fondly as the machine flashes its eyes green, a color meant to indicate happiness and excitement. His son mostly ignores it.

Yahini Nozue: We talked about buying it but we said, "Oh, it's very expensive." And my husband said, "Should we buy it or not?" I never seriously thought he would buy it, but he did [laughs]!

Faith: Do you enjoy it?

Yahini: As a pet I enjoy it. If it were really alive it would be troublesome. When I don't want it around, I can just put it back on the charger, so it's a good pet.

Do you play with it?

Mitsuhiko Nozue: We don't actually play with AIBO. It just walks around. He can walk by himself so he plays by himself.

What did you think when you got this?

Yahini: We were very nervous. We weren't sure how we should touch it.

You had to teach it, right? It just had basic skills when it came?

Mitsuhiko: There was a basic program—

Yahini: It could pee at first—no liquid comes out. It just cocked its leg and made a sound. That was in the basic program. But it didn't do a lot. The training is that if it goes and gets the ball, it plays with it—things like that. But we really don't use the instructions any more. It's trained.

How smart would you say it is?

Mitsuhiko: It's not very intelligent. It doesn't recognize who we are and it bumps into things.

So it walks into the furniture?

Mitsuhiko: Yes, and it can't go up even little steps [points to a slight ledge between living-room area and bedroom]. It finds it difficult to balance, as well.

Do you ever take it out of the house?

Yahini: The first time we ever took it out of the house is when we brought it to the AIBO fair.

There was an AIBO fair in Japan?

Mitsuhiko: Yes, everyone who bought AIBO could go, and because everyone applied to get them over the Internet, there were people from all over Japan.

Why is AIBO just sitting there now, doing nothing?

Yahini: It must be thinking.

Mitsuhiko: Some people at the fair were so interested in the AIBO that they said they wanted to open up the AIBO and look at the interface and see how it's programmed. And they just play with it continuously, but we're not like that at all.
I know of some researchers who did that to see how it operates.

We watch it walk across the floor. Its motors make mechanical sounds as it moves. Its ears are floppy, rubbery plastic but the rest of the robot is hard, silvered plastic. Seven-year-old Masahiko is at eye level with the dog, down on his hands and knees.
Do your friends come over to see it?

Masahiko: Yes, they come.
How often do you play with it?
Masahiko: Sometimes I play with it but not very much.
Is it like a real pet?
Masahiko: No.
I guess nothing replaces a real dog.
Yahini: He really has no interest in mechanical things as a pet.
[When Peter changes the light for his photography, the robot makes cheering-crowd noises.]
Masahiko: I wonder why it got happy all of a sudden.
Yahini: Maybe because there's more light.
Do you wish that it had more sensors?

Seven-year-old Masahiko Nozue gets down on the floor and romps with AIBO (above), Sony's robotic pet dog. The Nozues had wanted a real dog, but pets are not allowed in their apartment. AIBO never needs to be fed, bathed, or walked, although it can simulate urination; it doesn't shed hair, bark at the neighbors, or need to be kept in a kennel when its owners go on vacation. Still, its behavior is so lifelike that the Nozues find it hard to treat it like a machine. One charge on its rechargeable battery lasts about two hours, and during that time AIBO is for all intents and purposes one of the family.

The novelty of owning Japan's first robot dog is not enough to keep Mitsuhiko Nozue's son Masahiko from switching his attention to a Pokemon video game. When abandoned by its owner, AIBO—Sony's new, limited-edition mechanical pet—plays with the ball by itself, delighting Mitsuhiko. The man runs for the 150-page manual that came with the robot pet when AIBO displays any new trick, sometimes leaving Mitsuhiko scratching his head—a puzzlement all too familiar from other encounters with digital gizmos. The latest word is that the Nozue family has named their AIBO Narubo. Mitsuhiko writes, "Recently, he is selfish. He sometimes doesn't play with the pink ball." Mitsuihiko wonders if Narubo is "growing from child to adult."

Mitsuhiko: Yes, I wish that. Dogs react more to touch, so I wish that it had sensors all over its body, like a normal dog. And it can only go for two hours and then has to be put back up on the battery charger to recharge.

Do you miss it when it's not walking around?
Mitsuhiko: When it's playing, it's noisy. Zaka-zaka-zaka-zaka. I look at a TV show—zaka-zaka-zaka-zaka! Sometimes it bumps into the step between the living room and the bedroom—zaka-zaka-zaka-zaka! But it's part of my life.

Do you wish AIBO could go up that step?
Mitsuhiko: It would be good if it could just move around without any sort of assistance.

When do you have it running during the day?
Mitsuhiko: In the mornings, I am here from seven until nine. I put it on and it moves around by itself and then it goes back into the charger. If it crashes into a wall or something, I turn it around. Otherwise, it just walks around by itself. I don't fuss around with it—I just let it walk around. There probably are people who spend all their time fussing over it, but I don't.

This is odd. AIBO is resting on its stomach, but it's lifted all four paws up in the air and is flailing them around in circles.
Mitsuhiko: This is the first time he has ever done this!

Why do you think it's doing that?
Mitsuhiko: Maybe because it's been getting praised a lot.
Yahini: When you pat the head softly many times [as she had been], that's praising it a lot, so maybe that's why. It looks like it's sky diving.

Mr. Nozue, what did your friends at work think when you got this AIBO?
Mitsuhiko: They weren't interested in the actual AIBO, just in the fact that I bought one, because it was so expensive.

Would you like to have more robots in your house?
Mitsuhiko: No.
Yahini: There would be too many things to look after if we got another one.

What about a vacuum-cleaning robot?
Yahini: In America, maybe, but a Japanese home is much smaller. There's no room to walk around and do the cleaning.
Mitsuhiko: I think of those as machines—not as robots like this. Also, there is a huge manual that came with AIBO and I also have this [a large loose-leaf notebook out, filled with instruction manuals for the family's electronics]. No more! No more electronics [laughs]!

Tell me how you differentiate between robots and machines.

Mitsuhiko: I don't think AIBO is a machine. We don't want to always be operating him. We want him to do whatever he wants.

Do you feel like it's alive?

Mitsuhiko: That's really difficult to say. It has got a remote control and if we want to control it we can use the remote, but we just let it move around freely. The only time we put any kind of control over him is when we want him to sit, because he has to sit to be put on the charger.

It's walking around making sounds—do you know why?

Yahini: It wants to kick the ball. It turned around and wanted to kick it but it didn't know where it was.

Have you had to bring it to the vet?

Mitsuhiko: Its tail got bent and it couldn't move properly anymore. We sent it to Sony and they sent it back with a new tail.

Did you miss it?

Mitsuhiko: Mmmm, I didn't really miss it.

If you didn't have AIBO anymore would you be unhappy?

Mitsuhiko: I couldn't give it away to anyone. I couldn't part with it.

You're attached to it?

Mitsuhiko: It's hard to explain how I feel about it.

Yahini: It's not just a toy.

It's a pet?

Yahini: Yes.

Do you consider it part of your family?

Mitsuhiko: I don't like to think of dogs as a part of the family, because they are pets—even if it's a real dog.

But you regard AIBO as a pet.

Mitsuhiko: Yes. It's part of the family as a pet. I think it has fallen asleep. It's not moving anymore.

Yahini: If you shake it around, it wakes up.

Is there an indicator for how much battery power there is left?

Mitsuhiko: There's no indicator, but he puts his head down if he's getting low.

Is he getting low now?

Mitsuhiko: It was playing around all morning, so it is just going to sleep.

It just lifted its leg and pretended to pee on the floor. Very odd—does it do that often?

Mitsuhiko: Usually not this much.

Is there anything it doesn't do that you would like it to do?

Mitsuhiko: Most Japanese dogs you can teach to shake hands, but you can't teach that to AIBO. And it understands the shape of a human face but doesn't recognize the individual face. I would like it to differentiate among us. I would like it to know me.

Relaxing in his office at the Mechanical Engineering Lab in Tsukuba, Japan, Takanori Shibata (above) pats a derivitive product from his research: a robot cat named Tama. Shibata is a roboticist who studied with MIT robot guru Rodney Brooks before heading his own lab. Omron, a Japanese engineering company, applied Shibata's discoveries to produce Tama—a mechanical pet with sensors beneath its fur that react to sound and touch. If Shibata calls the cat's name, it hunches slightly in a welcoming posture, inviting a caress. Stroking the machine's back elicits a purr; loud noises cause it to startle and look around for the source of the sound. Five machines have been built thus far at half a million yen apiece (about $5,000). Omron says it has no plans as of yet to commercialize its robot cats.

Final Fantasy

Robo Specs

Name: BIT

Origin of name: IT was the earlier face robot. This is Baby IT.

Purpose: Prototype toy to try to get major toy manufacturers interested.

Creative inspiration: Wanted to make a toy that was more realistic than existing dolls.

Height: Baby size

Length: Baby size

Weight: Baby weight

Vision: None

Sensors: Orientation, reed switches, microphone, light sensor

Frame composition: Plush body, machined plastic box-frame head

Actuator type, number, and kind: 5 very cheap electric motors

Batteries: 8 AA, 1 9 V battery

Cost of components: Dirt cheap

Retail price of My Real Baby (production model of Baby IT prototype): $95

Project status: On sale Christmas 2000

Information from: Rodney Brooks, Colin Angle

Peter's Notes

The folks at iRobot winced when I asked them to cut the baby's face in half. They understood, however, that photographing their cute robot doll's expressions without seeing the mechanics involved just wouldn't cut it. So we did, with an X-acto knife. Dressed in a Baby Gap outfit, the half-faced doll was laid out on a cluttered workbench where it was nestled among tools, batteries, and some spare faces that were used to create this prototype. The photo took a few hours because the lighting was tricky. For some shots I used a focused spotlight to make the exposed eye glow yellow, red, or green. But the effect ended up being overkill—the straight photo was creepy enough—maybe too creepy. When I showed the picture to roboticist Matsuo Hirose in Tokyo (the chief engineer of the Honda humanoid project), he shuddered. "I don't like this picture," he said. "I feel like this baby might do something to me...."

Much thought went into the robot doll jointly developed by the Hasbro toy company and iRobot, the robotics firm co-founded by Rodney Brooks of the Massachusetts Institute of Technology (see page 58). When we went to visit the prototype at iRobot's headquarters (formerly IS Robotics) in a strip mall northwest of Boston, it was a hush-hush operation—they wanted to show off their creation, but not reveal anything important about it. At first, Brooks did not want us even to mention that he was working with Hasbro. Ultimately, he relented. We were introduced to the doll by Colin Angle, the chief executive officer of iRobot, and Helen Greiner, its president.

Introduced under the name My Real Baby in February 2000 by iRobot and Hasbro, the robot can imitate many of the behaviors of a human infant. Rather like a real newborn, My Real Baby chews, sucks, coos, and cries. When the doll's owner first picks it up, My Real Baby opens its eyes and smiles. It giggles when tickled, complains when it is "hungry," and, as Brooks showed us, cries when shaken violently by the leg. If it is rocked nurturingly, the crying stops. It has a bottle with a chip inside; put the bottle near the baby and it will chew and suck on the nipple. Over time, it "learns" simple phrases, telling its owner phrases like "I love you, Mama," "I want baba, Mama," and "Night, night."

In an attempt to depict what the robot is really like, Peter asked iRobot's Angle and Greiner if he could strip off some of the soft, rubbery skin. They agreed and Peter sliced off half of the robot's face with a Xacto knife. The resulting photograph was creepy—surprisingly so, given that I had always known that the inside of My Real Baby was filled with machinery. Why is it disturbing to see the gears I already knew existed? Given that we had already spent months looking at robots of every size and description, why am I bothered by this one?

The image directly evokes the fears about robots—about technology generally—that everyone shares, to one extent or another. With its face stripped away, the doll looked like a baby that had been replaced by a machine. That the machinery was usually hidden only added to the effect. It recalls the climaxes of the countless science-fiction films in which the treacherous android's real identity is revealed. The toy industry is the means by which technology is slipped into the home—a stealthy infiltration.

Since the Greeks, Western societies have been haunted by the thought that their vaunted technology might end up destroying them. Frankenstein helped to crystallize the image of the machine-made human—powered by electricity, overwhelmingly strong, erratically murderous. From the beginning, laments the heroine of *R.U.R.*, the Czech play that brought the word "robot" to the English language, "There was such a cruel strangeness between us and them." In the end, the robots storm the factory in which they were made and murder their creators. More than seven decades later, an essentially similar plot formed the backbone of the wildly successful Terminator films, both of which ended with the human-looking killer robot played by Arnold Schwarzenegger getting half of its face burned off to reveal the mechanism beneath.

The half-doll in iRobot's headquarters called on these free-floating anxieties—the notion that the friendly mask of technology would slip away, revealing an inimical force that would wipe out human society. How could it not? Far from being the hysteria of the technophobes, these fears have been given a loud voice by some of the world's most prominent technologists.

Two weeks after My Real Baby made its debut at the annual Toy Fair in Manhattan, Bill Joy, chief scientist of Sun Microsystems, made the front page of the *New York Times* by predicting that tiny, super-intelligent robots would inevitably learn to replicate themselves. "Once an intelligent robot exists," he warned, "it is only a small step to a robot species—to an intelligent robot that can make evolved copies of itself." And, argues nanotechnology visionary Eric Drexler, such replicating machines "could easily be too tough, small, and rapidly spread to stop." Humankind's "technical arrogance," the physicist Freeman Dyson has lamented, may be the ultimate cause of all its troubles. "Which is to be master?" Joy asked. "Will we survive our technologies?"

With dark visions such as these becoming common, it is little wonder that exposing the mechanical reality beneath the humanlike skin of the prototype robot doll was unsettling. Yet as we looked at the wires and actuators a second image took shape as well. Implicit in the fears of the technologists is the notion that people and machines are inevitably locked in struggle, and that people will inevitably lose. But that seems a little simplistic, doesn't it? The relationship between humankind and its creations is more complex than the simple Manichaean death-struggle envisioned by roboticists Kevin Warwick and Hans Moravec.

For every technology that oppresses us, there is another that expands the realm of hope. Hundreds of millions, if not billions, of people are alive today because of the innovations that led to antibiotics

and the Green Revolution. Pacemakers, artificial hips, robotic limbs, and fast, cheap, computerized diagnostic tests have relieved the suffering of countless multitudes. The examples could be multiplied a dozen-fold. In each case people capitalized, for better and worse, on aspects of the technology that they liked, ignored the rest, and tried to avoid the consequences, sometimes with amazing success.

It may well be the same for robots. Fifty years from now, says roboticist Hiroaki Kitano (see page 212), machines and humans will battle it out at the world RoboCup championship. But it is far more likely that Kitano's successors will have trouble figuring out who should belong on each side.

As robots become ever more socialized and people are ever more technologically or biotechnologically enhanced, the two sides may come to resemble each other more than they differ. Artificial hips lead to cochlear and retinal implants lead to implanted chips lead to—who knows?—depositing memories

in computers? People may work shoulder to metal shoulder with robots who have the recorded personalities of human beings.

The terrible fear, and great hope, is that we may lose some of our humanity. It may happen. With good luck, we might lose some of the poverty, fear, and desperation that has always been the human lot. With bad luck, we may eventually destroy ourselves. But in either case the RoboCup of 2050 begins not with humans and robots facing off, but with both sides standing together. The players, each a different combination of person and machine, shake hands. In the noise and glare of the crowd, one can imagine it being hard to tell them apart. They line up, indistinguishable from afar, on either side of the midfield line, metal and flesh, flesh and metal, all in an inextricable tangle.

The spectators' shout reaches a crescendo as the first foot touches the ball. It is the first whisper of evolution, the dawn of a new species: *Robo sapiens.*

Its skin partially removed to reveal its inner workings, this prototype robot baby can mimic the facial expressions of a human infant by changing the contours of its lifelike rubber face. Called BIT, for Baby IT, the mechanical tot is yet more proof that much robotic research will see its first commercial application in the toy and entertainment industry. My Real Baby, the market version of BIT, is scheduled to debut in US stores in late 2000; it is a collaboration between Hasbro, the US toy giant, and iRobot, a small company started by MIT researcher Rodney Brooks (see page 58). Sensors beneath the doll's skin respond to a child's touch, so the robot giggles when tickled and requests a bottle after a certain period of play. Colin Angle, CEO of iRobot, confidently asserts that My Real Baby will sell so much that the company "will own Christmas 2000."

Surrounded by his robot toys and sculptures, Clayton Bailey of Port Costa, California, is living proof that not everyone feels threatened by the prospect of being surrounded by mechanical human beings. In the studio behind his home, Bailey stands among the large robots he has sculpted since retiring as a professor of art from California State University, Hayward. He and his wife, Betty, have collected robot and space toys for the past 30 years.

Methodology

The procedure of inquiry we followed for this thinking person's book of photography began with choosing the projects on which to focus. Those we chose, over one hundred in all, were based on academic papers, the recommendations of other researchers, websites (though often out of date), lab visits, and personal preference. Our purpose in producing this book was not to create an encyclopedic look at robotics as a *tour d'horizon* of the field. Doubtless we missed some important projects; to their makers, we offer apologies and the hope that the general overview we provide is worth a few blurred details.

Peter and I consider our book design and production to be part of the editorial process, and therefore it remains organic until the last—a bitter pill if time runs late or the first chapter manuscript suddenly becomes a jumble of unintelligible corrupted documents.

For those interested in Peter's photographic *modus operandi*, he offers the following:

"Taking photos in labs, at least the kind of photos I like to take, disrupts research. Fortunately, all the researchers I photographed for this book understood (or at least they did after I finished) that making interesting photos of their work takes time. I used a lot of lighting to bring out the best in the robots, using everything from pencil light flash tubes to automotive spot lights. Most of the time I used a combination of five flash heads with various softboxes and gridspots powered by two Dynalight packs. Sometimes I used double exposures, but there is no digital manipulation of any image—all photos were made in camera. I used both 35mm Canon and 6 x 4.5 Mamiya cameras, shooting more than 200 rolls of color transparency film, both Fujichrome and Ektachrome. Indispensable tools include the Polaroid backs for both cameras, which provide instant workprints for fine-tuning the lighting and a well-developed sense of humor and patience, always useful, for dealing with SNAFU situations."

Though the robots in this book may have further evolved since the interviews and photographs herein, we decided early on to use the point in time at which an interview was done to stop the technological time clock.

Researchers and other interviewees were told in advance that the text would be conversational in nature and therefore allowed a tape recording to be made of our conversations during the reporting phase. In most cases, this record was supported by subsequent written communiques and clarifications. Though a very few researchers (some terrified after having made rather grandiloquent statements about their work and its impact) were appalled to see their own words in print, I thought it important for the reader that the individual's style of speaking be maintained. I was careful not to overemphasize points that were not, in fact, emphasized greatly by the scientist, engineer, or technologist. For more comprehensive treatment of the work of these researchers, I direct you to their academic papers, which are often available on the Internet.

As a rule, I don't pass manuscript back to those I've interviewed, but with this book, I decided it was a good idea to do so—in order to double-check the accuracy of my own understanding of the science and technology involved. Also, because question-and-answer text by its nature can be difficult to reduce to a readable but wholly accurate summary, this stage also helped to ensure the fairness of the conversational record. Most everyone worked within the very narrow constraints I had set for restating themselves and therefore contributed to making the text both indicative of their work and also interesting.

In the editing process I sometimes changed the wording of a question, either to provide contextual information or to provide a definition to help the reader understand the answer. Naturally, I was careful not to change the meaning of my questions.

Researchers for whom English is not their first language were normally offered the assistance of a translator for the interviews. In most cases, scientists and technologists, who speak English as a second or third language, are quite comfortable with it and preferred using it to having a translator present.

In some cases I was unable to include an interview because of the space constraints dictated by the photography in the book. Although my conversations with several people informed the rest of the book, I could not use these interviews directly. For this reason, my gratitude and apologies to

Robin Murphy, Ken Salisbury, John Hollerbach, Dan Koditschek, Johann Borenstein, and Mark Cutkosky. Of necessity, not only serious science and research concerns but other revealing moments often were relegated to the cutting room floor. For example:

The University of Michigan's Johann Borenstein, the inventor of the guidecane which is a helpful extension of a blind person's senses, laments the lack of funding for his research—a problem, it seems, of the economies of scale:

Faith: The guidecane made you famous, huh Johann?

Johann Borenstein: Yeah, it sure did. It won a *Discover* magazine award. It was in papers and magazines all over the world. It's too bad that the promotion and stuff didn't result in more funding. That's really ticking me off because it could really be a good product once we refine it. There would have to be more blind people in order for it to be funded.

In addition to his all-too-common researcher's angst, Borenstein has obviously been smitten by what he refers to as the cruel realities of life:

Borenstein: Asimov had a trilogy, I think one of them is *I, Robot*, where there is the concept presented of a woman living together with what we would call a robotic husband.

Faith: Gladia in Robots of Dawn. *It was fascinating.*

I have taken that concept to many dinner parties. The question that I pose is, if you could order at the store the perfect robot—a wife in my case. Would you to do that? Suppose I could get it perfect in every way, completely to my specifications, would I want to do that?

Would you?

Yes. Absolutely. Any time!

Have you ever seen The Stepford Wives? *It's not a great movie, but it's about a community where all the men get together and rebuild their wives, replace them with robots.*

Oh, a dreamworld.

That's fairly appalling, Johann.

I guess I am tainted by the cruel realities of life.

Are women the cruel realities of life?

I'll tell you what: I think you could sum it up in the "All Important Off Button."

I believe we should move on to another topic.

Even a Pause button would be———

Charles C. Mann, our editor, managed to weed out the extraneous and ill-advised, keep us focused, and lend a hand at writing in the all-too-common pinches. We also relied on the visual and editorial acumen of our designer David Griffin. The light, precise hand of copyeditor Katherine Wright is invisible throughout.

I found the following books invaluable to my research:

The Age of Spiritual Machines by Raymond Kurzweil

Brainmakers: How Scientists are Moving Beyond Computers to Create a Rival to the Human Brain by David H. Freedman

Darwin's Dangerous Idea: Evolution and the Meanings of Life by Daniel C. Dennett

The End of Science by John Horgan

Inside the Robot Kingdom by Frederick L. Schodt

Mind, Language and Society: Philosophy in the Real World by John R. Searle

The MIT Encyclopedia of the Cognitive Sciences (Robert A. Wilson and Frank C. Keil, editors)

Nonzero: The Logic of Human Destiny by Robert Wright

The Society of Mind by Marvin Minsky

The Technological Society by Jacques Ellul

— Faith D'Aluisio

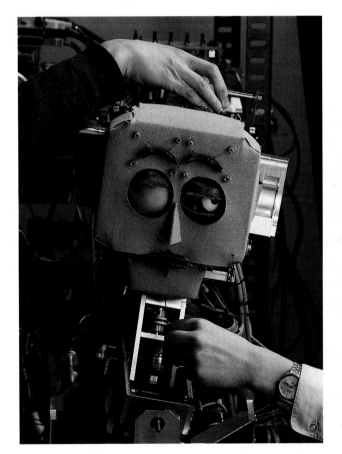

Researchers adjust the mechanism of WE-3RIII, Waseda University's head robot, after it accidentally whiplashed into its own wires.

Glossary

A priori knowledge: Facts or ideas relating to or derived by reasoning from self-evident propositions.

Actuator: A motor. A transducer that uses electrical, hydraulic, or pneumatic energy to make a robot move.

Algorithm: A formula or set of steps for solving a particular problem. To be an algorithm, a set of rules must be unambiguous and have a clear stopping point.

Analog: A quantity that is continuously varying, as opposed to varying in discrete steps. Contrasted with digital.

Animatronic: Being or consisting of a lifelike electromechanical figure of a person or animal that has synchronized movement and sound.

Android: A robot that approximates a human in physical appearance. A humanoid.

Anthropomorphism: The attributing of human characteristics to non-human entities.

Architecture: The physical and logical structure of a computer or manufacturing process.

Artificial Intelligence (AI): The mode of programming that allows a computer to operate on its own; for example, to learn, adapt, reason, correct, or improve itself.

Artificial life: Simulated organisms, each including a set of behavior and reproduction rules (a simulated "genetic code") and a simulated environment. In a computer, the simulated organisms go through multiple generations of "life" and "evolve." The term can refer to any self-replicating pattern.

Automaton: A mechanism that is relatively self-operating, like a robot, and designed to follow automatically a predetermined sequence of operations, or respond to encoded instructions.

Autonomous: Existing or capable of existing independently.

Bandwidth: A communication term that refers to the speed and carrying capacity of a communication channel.

Behavior-based robotics (BBR): A methodology that bridges the fields of artificial intelligence, engineering, and cognitive science in order to design autonomous agents and robots. Its goal is to develop methods for controlling artificial systems and to use robotics to model and better understand biological systems.

Bioengineering: The field that designs strains of plant and animal life by directly modifying the genetic code.

Biology: The study of life forms. In evolutionary terms, the emergence of patterns of matter and energy that could survive and replicate to form future generations.

Biomimesis: To mimic life, to imitate biological systems.

Bit: A contraction of the phrase "binary digit." In a binary code, one of two possible values, usually zero and one. In information theory, the fundamental unit of information.

Bootstrap: A technique for loading the first few instructions of a routine into storage, then using these instructions to bring the rest of the routine into the computer from an input device. This usually involves either the entering of a few instructions manually or the use of a special key on the console.

Brachiation: In animal behavior, the specialized form of arboreal locomotion in which movement is accomplished by swinging from one hold to another by the arms, as the way monkeys use their arms to travel through the trees.

Bus: One or more conductors used for transmitting signals or power. An information coding scheme by which different signals can be coded and identified when sharing a common data channel.

Byte: A contraction for "by eight." A group of eight bits that stores one unit of information on a computer. A byte may correspond, for example, to a letter of the alphabet.

CAD: Computer-aided design.

CCD camera: A solid-state camera that uses a CCD (charge-coupled device—an electonic imaging device containing layers of silicon that release electrons when struck by light) to transform a light image into a digitized image.

CD-ROM: Stands for "compact disc read-only memory." A laser-read disc that contains up to a half billion bytes of information. "Read only" refers to the fact that information can be read, but not deleted or recorded, on the disc.

Chaos: The amount of disorder or unpredictable behavior in a system.

Closed loop: A programming loop that has no exit and whose execution can be interrupted only by intervention from outside the software in which the loop is included.

Closed system: Interacting entities and forces not subject to outside influence (for example, the universe).

Cochlear implant: A surgical implant that performs frequency analyses of sound waves similar to that performed by the inner ear, in an attempt to reproduce the experience of hearing.

Colossus: The first electric computer, built by the British from 1500 radio tubes during World War II. Colossus and nine similar machines running in parallel cracked increasingly complex German military intelligence codes and contributed to the Allied victory.

Computational neuroscience: Understanding in computational terms how the brain generates behaviors.

Computer: A machine that implements an algorithm. A computer transforms data according to the specifications of an algorithm. A programmable computer allows the algorithm to be changed.

Computer language: A set of rules and specifications for describing an algorithm or process on a computer.

Consciousness: The ability to have subjective experience. The ability of a being, animal, or entity to have self-perception and self-awareness.

Control: The process of making a variable, or a system of variables, conform to what is desired. A device to achieve such conformance automatically.

Controller: An information-processing device whose inputs are both the desired and the measured position, velocity, or other pertinent variables in a process and whose outputs are drive signals to a controlling motor or actuator.

Cybernetics: A term coined by scientist Norbert Wiener to describe the "science of control and communication in animals and machines." Cybernetics is based on the theory that intelligent living beings adapt to their environments and accomplish objectives primarily by reacting to feedback from their surroundings.

Cyborg: A bionic human. A human having normal biological capability or performance enhanced by or as if by electronic or electromechanical devices.

Database: A structured collection of data that is designed in connection with an information-retrieval system.

Debugging: The process of discovering and correcting errors in computer hardware and software.

Deep Blue: A computer program, created by IBM, that defeated Gary Kasparov, the world's chess champion, in 1997.

Degree of freedom: One of a limited number of ways in which a point or a body may move or in which a dynamic system may change, each way being expressed by an independent variable and all required to be specified if the physical state of the body or system is to be completely defined.

Digital: Varying in discrete steps. The use of combinations of bits to represent data in computation. Contrasted with analog.

Effector: An actuator, motor, or driven mechanical device.

Endoscope: A flexible instrument for seeing inside organs of the human body.

Entropy: A measure of the unavailable energy in a closed thermodynamic system, which is usually also considered to be a measure of the system's disorder. Also, the degradation of the matter and energy in the universe to an ultimate state of inert uniformity; a process of degradation or running down or a trend to disorder.

Ethology: The scientific and objective study of animal behavior, especially in natural conditions.

Evolution: The historical development of a biological group (as a race or species). A theory that the various types of animals and plants have their origin in the other preexisting types and that the distinguishable differences are due to modifications in successive generations.

Evolutionary algorithm: Computer-based, problem-solving systems that use computational models of the mechanisms of evolution as key elements in their design.

Feedback loop: The components and processes involved in correcting or controlling a system by using part of the output as input. A loop is a sequence of instructions that is executed repeatedly until some specified condition is met.

Field robotics: The use of mobile robots in field environments, such as work sites and natural terrain, where the robots must safeguard themselves while performing nonrepetitive tasks and objective sensing as well as self-navigation in random or dynamic environments.

Force-feedback: A sensing technique using electrical or hydraulic signals sent back to a human operator controlling a robot end-effector, or tool.

Fovea: An area at the center of the retina where the cells that see color (cones) are concentrated and there are no cells that detect dim light (rods).

Genetic algorithm: A model of machine learning that derives its behavior from a metaphor of the mechanisms of evolution in nature. Within a program, a population of simulated individuals is created and undergoes a process of evolution in a simulated competitive environment.

Genome: The genetic material of an organism.

Haptic feedback: Feedback relating to or based on the sense of touch.

Homo sapiens: Human species that emerged perhaps 400,000 years ago. *Homo sapiens* are similar to advanced primates in terms of their genetic heritage and are distinguished by their creation of technology, including art and language.

Humanoid: In robotics, a robot that approximates a human in physical appearance. An android.

Inclinometer: An instrument for indicating the inclination to the horizontal axis (as of an airplane).

Iteration: The repetition of a sequence of computer instructions a specified number of times or until a condition is met.

Kinematics: The branch of mechanics concerned with the motions of objects without being concerned with the forces that cause the motion. In this latter respect it differs from dynamics, which is concerned with the forces that affect motion.

Kinesthesia: The awareness of one's own body parts, weight, and movement mediated by organs located in muscles, tendons, and joints.

KLOC: 1,000 lines of code in a computer program. (K=1,000, LOC=lines of code).

Laparascope: A flexible fiber-optic instrument inserted through an incision to visually examine the interior of the body.

Luddite: One of a group of early-nineteenth-century English workmen who destroyed labor-saving machinery in protest. The Luddites were the first organized movement to oppose the mechanized technology of the Industrial Revolution. Today, the term more broadly refers to opposition to technology.

Machine learning: Computer systems that can adapt and learn from their experience.

Manipulator: A flexible mechanical device that performs certain motions or tasks automatically, such as picking cartons off a series of pallets and then putting them on a conveyor belt to be put into a truck.

Mechatronics: A term coined in Japan in the 1970s: the fusion of machinery and electronics. Intelligent mechanisms.

Microprocessor: An integrated circuit built on a single chip containing the entire central processing unit (CPU) of a computer.

Moore's Law: Prediction by engineer Gordon Moore that the number of components on a representative chip will double every one to two years.

Morphology: A branch of biology that deals with the form and structure of animals and plants. The form and structure of an organism or any of its parts.

Nanoengineering: The design and manufacture of products and other objects based on the manipulation of atoms and molecules; building machines atom by atom. "Nano" refers to a billionth of a meter, which is approximately the width of five carbon atoms.

Nanosecond: One-billionth of a second.

Nanotechnology: The building of devices on a molecular scale.

Nanotubes: Elongated carbon molecules that resemble long tubes, formed of the same pentagonal patterns of carbon atoms as buckyballs (soccer-ball-shaped molecules). Nanotubes may be able to perform the electronic functions of silicon-based components.

Neural implant: A brain implant that enhances one's sensory ability, memory, or intelligence.

Neural network: A type of artificial-intelligence model that attempts to imitate the way the human brain works. Rather than using a digital model, in which all computations manipulate zeros and ones, a neural network works by creating connections between processing elements, the computer equivalent of neurons. The organization and strengths of the connections determine the output. A computer simulation of human neurons.

Neuron: The fundamental functional unit of nervous tissue.

Pneumatic: Moved or worked by air pressure.

Pick-and-place robot: A simple robot, often with only two or three degrees of freedom, that transfers items from one place to another by means of point-to-point moves.

Piezoelectric reaction: The electric current or electric polarity produced by applying pressure to a crystalline substance, such as quartz.

Polymer: A chemical compound or mixture of compounds in which two or more molecules combine to form larger molecules (the polymer) that consist of repeating structural units.

Potentiometer: A voltage divider. A resistor or a chain of resistors connected in series that can be tapped at one or more points to ascertain the total voltage across the whole resistor or chain.

Program: A set of instructions that enables a computer to perform a specific task.

Quantum computing: A potentially revolutionary method of computing, based on quantum physics, that uses the ability of subatomic particles such as electrons to exist in more than one state at a time.

Quantum mechanics: A theory that describes the interactions of subatomic particles, combining several basic discoveries.

Random Access Memory (RAM): Memory space in a computer where the user can alter and temporarily store information.

Robo sapiens: A hybrid species of human and robot with intelligence vastly superior to that of purely biological mankind; will begin to emerge in the twenty-first century.

Robot: A machine that looks like a human being and performs various complex acts (as walking and talking) of a human being. A mechanism guided by automatic controls.

Robot platform: A robot that is made with standard production features and is used for research.

Self-replication: A process by which an organism or device creates a copy of itself.

Semiconductor: A material commonly based on silicon or germanium with a conductivity midway between that of a good conductor and an insulator. The unusual electrical properties of semiconductors are the basis for the transistor, the fundamental component of a computer chip.

Sensor: A transducer or other device whose input is a physical phenomenon and whose output is a quantitative measure of that physical phenomenon.

Servo: Also servomechanism or servomotor. An automatic control mechanism consisting of a motor driven by a signal that is a function of or the difference between a commanded position and/or rate and an actual measured position and/or rate.

Glossary CONTINUED

Shape deposition manufacturing: The embedding of electronic parts into molded plastic to create structures with the flexibility of living tissue.

Shape-memory alloy: An alloy (a metal mixed with other metals or non-metals to give it special qualities) that has the ability to regain its shape after it has been bent. (Usually mixtures of nickel, titanium and copper.)

Simulacrum: A faint, shadowy, or unreal likeness; a mere semblence.

Solid-state: Electronic component consisting entirely of solids (semi-conductors, transistors) rather than electron vacuum tubes.

Subsume: To include or place within something larger or more comprehensive.

Subsumption architecture: A layered approach to assembling reactive rules into complete control systems from the bottom up. Each layer gives the system a set of prewired behaviors. The higher levels build upon the lower levels to create more complex behaviors. The behavior of the system as a whole is the result of many interacting simple behaviors. The layers operate in such a way that there is no timing requirement for transmission and the start of each character is individually signaled by the transmitting device.

Technology: The practical application of knowledge, especially in a particular area. An evolving process of tool creation to shape and control the environment. It requires invention and is itself a continuation of evolution by other means.

Tele-operation: A "master-slave" system of manipulation in which a robot autonomously performs repetitive or predictable actions that have been programmed in advance while under the supervision of an operator in a remote location, who intervenes when the robot must perform complicated tasks.

Transducer: A device that is actuated by power from one system and supplies power, usually in another form, to a second system (a loudspeaker is a transducer that transforms electrical signals into sound energy).

Turing machine: A simple abstract model of a computing machine, designed by Alan Turing in his 1936 paper "On Computable Numbers."

Turing test: A procedure proposed by Alan Turing in 1950 for determining whether or not a system (generally a computer) has achieved human-level intelligence, based on whether it can deceive a human interrogator into believing that it is human.

Vestibulo-ocular reflex: A normal reflex in which eye movement compensates for movement of the head. It is caused by stimulation of the inner ear structures (vestibular apparatus) that are associated with balance and position sense.

Virtual reality: A simulated environment in which you can immerse yourself. A virtual reality environment provides a convincing replacement of the visual and auditory senses.

Yaw: An angular displacement left or right viewed from along the principal axis of a body having a top side, especially along its line of motion.

Zero-Moment Point (ZMP): The point of intersection, with the ground, of the resultant of the forces acting on a bipedal robot. For dynamically-stable motion of a legged machine the trajectory of the zero-moment point should lie within the polygon of support.

Concise Science Dictionary. Oxford, UK: Oxford University Press, 1991.

Encyclopedia Britannica, http://www.eb.com.

Field Robotics Webpage of Robotics Institute of Carnegie Mellon University, http://www.frc.ri.cmu.edu.

Howe, Dennis. *The Online Computing Dictionary.* 1999. http://www.nightflight.com/foldoc.

Kohl, Herbert. *From Archetype to Zeitgeist.* New York: Little, Brown and Company, 1992.

Kurzweil, Ray. *The Age of Spiritual Machines.* New York: Viking, 1999.

Langone, John. *National Geographic's How Things Work: Everyday Technology Explained.* Washington, DC: National Geographic, 1999.

Merriam Webster's Collegiate Dictionary, Electronic Edition. Springfield, MA: Merriam-Webster, 1994.

Mosby Medical Encyclopedia, Revised Edition. New York: SIGNET, 1996.

Schodt, Frederik. *Inside the Robot Kingdom.* Toyko & New York: Kodansha International, 1988.

Takanishi, Atsuo. "Humanoid Robots: A New Tide towards the Next Century of Natural Human-Robot Collaboration." Department of Mechanical Engineering, Waseda University, Tokyo, Japan, 1998.

The Way Science Works. New York: Macmillan, 1995.

Tver, David F., and Roger W. Bolz. *Robotics Sourcebook and Dictionary.* New York: Industrial Press, 1983.

Webopedia, http://www.webopedia.com.

Webster's New Universal Unabridged Dictionary. New York: Simon & Schuster, 1983.

Wilson, Robert A., and Frank C. Keil, eds. *The MIT Encyclopedia of the Cognitive Sciences.* Cambridge, MA: The MIT Press, 1999.

World Book CD ROM. Chicago: World Book, 1998.

Recommended Reading

NON-FICTION

Arkin, Ronald C. *Behavior-Based Robotics (Intelligent Robots and Autonomous Agents).* Cambridge, MA: The MIT Press, 1998.

Brooks, Rodney A., and Anita Flynn. "Fast, Cheap and Out of Control: A Robot Invasion of the Solar System." *Journal of the British Interplanetary Society* 42, 1989.

Brooks, Rodney A. *Cambrian Intelligence: The Early History of the New AI.* Cambridge, MA: The MIT Press, 1999.

Darwin, Charles. *The Origin of Species.* Reprint. New York: Mentor, 1958.

Dennett, Daniel C. *Darwin's Dangerous Idea: Evolution and the Meanings of Life.* New York: Simon and Schuster, 1995.

Drexler, K. Eric. *Engines of Creation.* New York: Doubleday, 1986.

Dyson, Freeman. *From Eros to Gaia.* New York: HarperCollins, 1990.

Dyson, George B. *Darwin among the Machines: The Evolution of Global Intelligence.* Reading, MA: Perseus Books, 1997.

Feynman, Richard P. *The Feynman Lectures in Physics.* Reading, MA: Addison-Wesley, 1965.

Freedman, David H. *Brainmakers: How Scientist Are Moving Beyond Computers to Create a Rival to the Human Brain.* New York: Simon and Schuster, 1994.

Gell-Mann, Murray. *The Quark and the Jaguar: Adventures in the Simple and Complex.* New York: W.H. Freeman, 1994.

Hanley, Richard. *Is Data Human?* London: Boxtree, 1977.

Hofstadter, Douglas R., and Daniel C. Dennett. *The Mind's I: Fantasies and Reflections on Self and Soul.* New York: Basic Books, 1981.

Hofstadter, Douglas R. *Gödel, Escher, Bach: An Eternal Braid.* New York: Basic Books, 1979.

Krauss, Lawrence M. *The Physics of Star Trek.* New York: Harper Perennial, 1996.

Kurzweil, Raymond. *The Age of Intelligent Machines.* Cambridge, MA: The MIT Press, 1990.

———*The Age of Spiritual Machines.* New York: Viking, 1999.

Langton, Christopher G., ed. *Artificial Life: An Overview.* Cambridge, MA: The MIT Press, 1997.

Minsky, Marvin. *The Society of Mind.* New York: Simon and Schuster, 1985.

Moravec, Hans. *Mind Children: The Future of Robot and Human Intelligence.* Cambridge, MA: Harvard University Press, 1988.

———*Robot: Mere Machine to Transcendent Mind.* Oxford: Oxford University Press, 1998.

Mori, Masahiro. *The Buddha in the Robot.* Tokyo: Kosei Publishing, 1981.

Paul, Gregory S. *Beyond Humanity: Cyberevolution and Future Minds.* Rockland, MA: Charles River Media, 1996.

Penrose, Roger. *The Emperor's New Mind: Concerning Computers, Minds, and the Laws of Physics.* Oxford: Oxford University Press, 1989.

———*Shadows of the Mind.* Oxford: Oxford University Press, 1994.

Powers, Richard. *Galatea 2.2.* New York: Farrar, Straus, and Giroux, 1995.

Raibert, Marc H. *Legged Robots That Balance.* Cambridge, MA: The MIT Press, 1986.

Sagan, Carl. *The Dragons of Eden: Speculations on the Evolution of Human Intelligence.* New York: Ballantine Books, 1997.

Searle, John R. *Mind, Language and Society: Philosophy in the Real World.* New York: Basic Books, 1998.

Shodt, Frederik L. *Inside the Robot Kingdom.* Tokyo and New York: Kodansha International Ltd., 1988.

Stork, David G., ed. *HAL's Legacy: 2001's Computer as Dream and Reality.* Cambridge, MA: The MIT Press, 1996.

Wilson, Robert A., and Frank C. Keil, eds. *The MIT Encyclopedia of the Cognitive Sciences.* Cambridge, MA: The MIT Press, 1999.

FICTION

Asimov, Isaac. *I, Robot.* New York: Doubleday, 1950.

———*The Naked Sun.* New York: Doubleday, 1965.

———*Robot Dreams.* New York: Berkley Books, 1986.

———*Robots and Empire.* New York: Doubleday, 1985.

———*The Robots of Dawn.* New York: Doubleday, 1983.

Bradbury, Ray. *The Martian Chronicles.* New York: Doubleday, 1950.

Čapek, Karel, and Josef Čapek. *R.U.R. and The Insect Play.* New York: Oxford University Press, 1961.

Clarke, Arthur C. *3001: The Final Odyssey.* New York: Ballantine Books, 1997.

Huxley, Aldous. *Brave New World.* New York: Harper, 1946.

Shelley, Mary. *Frankenstein.* New York: Dover Publications (reprint), 1994.

Stephenson, Neal. *The Diamond Age.* New York: Bantam, 1995.

———*Snow Crash.* New York: Bantam, 1992.

Sterling, Bruce. *Schismatrix Plus: Includes Schismatric and Selected Stories from Crystal Express.* Reprint. New York: Ace Books, 1996.

Vinge, Vernor. *A Fire Upon the Deep.* New York: Tom Doherty Associates, 1992.

Index

AeroVironment, Inc., 156
  Black Widow, 156, 158
  DeMarino, Tom, 158
  MacCready, Paul, 27, 156, 157
  Mars Glider, 158
  Miralles, Carlos, 158
  Newbern, Scott, 157
  Spadaro, Marty, 158
  Trist, Paul Jr., 158
ALH84001 (meteorite), 140
Alien, 200
Ambrose, Robert J.(see Johnson
  Space Center)
American cockroach, 91
Ames Research Center, 124
Amidi, Omead (see CMU)
Anzen Taro, 171
Arizona centipede, 94
Arkin, Ronald C. (see Georgia
  Tech)
Asimov, Isaac, 37, 186
  I, Robot, 96
ATR (Advanced
  Telecommunications Research
  Institute Lab)
  DB, 50, 51, 52, 54, 55
  Dynamic Brain Project, 50
  Kotosaka, Shin'ya, 54
  Kawato, Mitsuo, 50, 52, 54, 55
  Schaal, Stefan, 50, 51, 54, 55
  Shibata, Tomohiro, 50, 52, 54
AutoDesk, 98
AVG, Inc., 207, 208
  Crypt Keeper, 207, 208
Ayers, Joe (see Northeastern
  University)

Bailey, Clayton, 231
Bailey, Sean (see Stanford
  University)
Baumgartner, Eric (see JPL)
Beer, Randal (see Georgia Tech)
Bellagio Casino Hotel, 217
Bennett, Drew (see iRobot)
Biosphere II, 114
Blade Runner, 114, 220
Body Snatchers, 219
Borchert, Thomas, 135
Boston Dynamics Inc., 24, 29
Breazeal, Cynthia (see MIT)
Brooks, Rodney (see MIT)
Buehler, Martin (see University of
  Michigan)
Burning Man Festival, 199

Čapek, Karel
  R.U.R., 23, 165, 228
Cardinal Health, Inc., 186
Carnegie Mellon University
  (CMU), 32, 161, 212
  Achim, Sorin, 214
  Ambler, 138
  Amidi, Omead, 160, 161
  Berkelman, Peter, 136
  Bowling, Michael, 214
  Center for Medical Robotics and
    Computer-Assisted Surgery,
    177
  CMUnited, 214
  Dante 1, 139
  Field Robotics Center, 138
  Kanade, Takeo, 23, 160

Moravec, Hans, 32, 228
Nomad, 138
Nourbakhsh, Illah R., 220, 221
Pioneer, 138
Robotics Institute, 23, 136, 145,
  160
  Shamah, Ben, 141
  Stone, Peter, 214, 215
  Veloso, Manuela, 214, 215
  Whittaker, William L. "Red," 138,
    139, 140, 141
Carnegie Museum of Natural
  History, 220, 221
  Sweet Lips, 220, 221
Case Western Reserve University,
  86, 96, 207
  Bachmann, Richard, 102
  Biorobotics Laboratory, 103
  Nelson, Gabe, 105
  Ritzmann, Roy, 102, 103, 104
  Robot I, II, III, IV, 102, 103, 104,
    105
  Watson, James, 102, 103, 104,
    105
  Quinn, Roger, 102, 103, 104, 105
Cham, Jorge (see Stanford
  University)
Cheng, Gordon (see Tsukuba
  Electrotechnical Laboratory)
Chernobyl, 138, 141
coelacanth, 107
Cog (see MIT)
Columbia University, 199
  Nayar, Shree, 145
Cray supercomputer, 129
Cutkosky, Mark (see Stanford
  University)

da Vinci, Leonardo, 217
Danbury Hospital, 186
Dario, Paolo, 39
DARPA (Defense Advanced
  Research Projects Agency), 27,
  101, 110, 111, 113, 155
de Garis, Hugo, 28, 32
Devol, George, 164, 165, 188
Diftler, Ron (see Johnson Space
  Center)
DiGioia, Anthony M., 177
Dilworth, Peter (see MIT)
Dinamation, 17
DLR (German Aerospace Research
  Agency)
  DLR Hand I, II, 5, 135
  Fischer, Max, 135
  ROTEX, 135
Draper Laboratories, 108
Drexler, Eric, 228
Dyson, Freeman, 228

Electrolux Robotic Vacuum
  Cleaner, 163
Engelberger, Joseph F., 23, 24, 164,
  165, 186, 188, 189
ERATO (Exploratory Research for
  Advanced Technology), 52, 80,
  213

FAO Schwarz, 1
Falk, Volkmar (see Intuitive
  Surgical)
Fiat, 188

FIDO (Field Integrated Design and
  Operations), 124, 127
Fischer, Max (see DLR German
  Aerospace Research Agency)
Forbes magazine, 98
Fujita, Yoshihiro (see NEC Corp.)
Fukuda, Toshio (see University of
  Nagoya)
Full, Robert J. (see University of
  California at Berkeley)
Furusho, Junji (see Osaka
  University)

General Electric, 188
General Motors, 188
  IRB 6400, 190
  Pontiac, Michigan Truck and
    Coach Plant, 190
Georgia Institute of Technology,
  152, 153
  Arkin, Ronald C., 27, 152, 153
  Atkeson, Christopher, 50
  Beer, Randal, 104
GMD (German National Research
  Center), 113.
  Institute for Autonomous
    Intelligent Systems, 195
  Kirchner, Frank, 27, 113, 195
  Kurt I, Kurt II, 195
  Schöll, Peter, 215
  Sir Arthur, 113
  Wilberg, Jörg, 215
Goodman, Bob, 179
Gray's paradox, 107, 108

Hara, Fumio (see Science
  University of Tokyo)
Harvard University, 98
Hasbro, Inc., 228
  My Real Baby, 12, 228, 229
Hawkes, Graham, 150, 151
  Deep Flight, 150
  Trap-T2, 150, 151
Helpmate Robotics, 186, 189
Herr, Hugh (see MIT)
Hirose, Masato, (see Honda Motor
  Company)
Hirose, Shigeo (see Tokyo Institute
  of Technology)
Hitchcock, Alfred, 73
Hollerbach, John M. (see
  University of Utah)
Honda Motor Company, 37, 48, 56,
  82, 152, 153, 196
  Hirose, Masato, 43, 45, 228
  P3, 35, 38, 43, 44, 45, 79, 183
Huntsberger, Terry (see JPL)

Ida, Fumio (see Science University
  of Tokyo)
IPA (Institut Produktionstechnik
  und Automatisierung), 195
Industrial Light and Magic, 200
Inoue, Hirochika, (see University of
  Tokyo)
Intuitive Surgical, 177
  da Vinci, 7, 175
  EndoWrist, 175
  Falk, Volkmar, 175, 176
  Shah, Sheila, 177
iRobot, 12, 19, 61, 64, 90, 109, 228,
  229

Angle, Colin, 61, 69, 228, 229
Ariel, 84, 101
Baby IT, 12, 19, 229
Bennett, Drew, 147
DiPietro, Alan, 86, 93, 147
Frost, Tom, 147
Greiner, Helen, 228
Urbie, 147
Williams, Ed, 101
Ivermectin, 219

Jacobsen, Steve (see SARCOS)
Japanese National Humanoid
  Project, 76
The Jetsons, 120
Johns Hopkins University
  Reza Shadmehr, 98
Johnson Space Center, 129, 130, 131
  Aldridge, Hal A., 133
  Ambrose, Robert J., 129, 130
  DART, 130, 131
  Diftler, Ron, 130
  Rehnmark, Fredrik L., 133
  Robonaut, 129, 130, 131, 133
Joy, Bill, 29, 30, 32, 228
JPL Jet Propulsion Laboratory, 147
  Aghazarian, Hrand, 126
  Baumgartner, Eric, 124, 126, 127
  Garrett, Mike, 124, 126
  Huntsberger, Terry, 124, 126, 127
  Kennedy, Bret, 127
  Nanorover, 127
  Rocky 7, 123, 124, 126
  Mars Rover, 124, 126, 127
  Thompson, Art, 127

Kanade, Takeo (see CMU)
Kandinsky, Wassily, 66
Kärcher, 165
Kato, Ichiro, 37, 38, 39
Kawato, Mitsuo, (see ATR)
Kirchner, Frank (see GMD)
Kismet, (see MIT)
Kitano Symbiotic Systems Project,
  212, 213
  Kitano, Hiroaki, 80, 82, 83
  Matsui, Tatsuya, 82, 83
  SIG, 80, 82, 83
  Symbiotic Intelligence Group, 80
Kodiara, Norio (see Mitsubishi)
Koditschek, Dan (see University of
  Michigan)
Kumph, John (see MIT)
Kuniyoshi, Yasuo (see Tsukuba)
Kurzweil, Ray, 29

Leg Labratory (see MIT)
Lego Mindstorm, 223
Los Alamos National Laboratory,
  24, 86, 117, 120

MacCready, Paul (see
  AeroVironment)
Marx, Groucho, 91
Massachusetts Institute of
  Technology (MIT), 25, 31, 57, 120
  Artificial Intelligence Laboratory,
    58, 62, 65, 66, 183, 184
  Aryananda, Lijin, 66
  Attila, 61
  Breazeal, Cynthia, 61, 66, 68, 69,
    70, 71, 73

Brooks, Rodney, 25, 39, 57, 58, 61, 62, 64, 65, 66, 69, 101, 120, 143, 227, 228, 229
  Cog, 25, 54, 55, 58, 61, 62, 64, 65, 66, 120
  Dilworth, Peter, 114, 115
  Fitzpatrick, Paul, 66
  Hawley, Michael, 200
  Herr, Hugh, 180
  Kismet, 66, 68, 69, 70, 71, 73
  Kumph, John, 108, 109
  Leg Laboratory, 8, 24, 86, 114, 115, 180, 183, 184
  Media Laboratory, 200
  Minsky, Marvin, 24, 31
  Paluska, Dan, 183
  Pratt, Jerry, 184
  Scassellati, Brian, 58, 61, 64, 65, 66
  Spano, Joseph, 180
  Spring Flamingo, 8, 184
  Troody, 114, 115
  Wanda, 108, 109
Masahiro, Mori
  Buddha in the Robot, 196
Material World, 224
McCarthy, John, 24
McGill University
  RHex, 96
Minsky, Marvin (see MIT)
MITI (Japanese Ministry of International Trade and Industry), 57
Mitsubishi
  Kodaira, Norio, 196
  Mitsubishi Heavy Industries, 107
  Terada, Yuuzi, 107
Mobot, Inc., 220
Modern Times, 200
Moravec, Hans (see CMU)
Morris, Errol
  Fast, Cheap, and Out of Control, 64
Murphy, Robin R. (see University of South Florida)

NASA, 109, 123, 127, 131, 145, 161, 168, 205
  Challenger, 140
  Mars Pathfinder mission, 58
  Sojourner, 58, 124, 138
  Space Shuttle, 129, 133
  near-Earth asteroid 4660
    Nereus, 127
NEC Corporation
  Bridgman, Aston, 166
  NEC Corporate Research Center, 166
  Fujita, Yoshihiro, 166, 167
  R100, 166, 167
New York Times, 228
Nomad (see CMU)
Nomadic Technologies Group, 152
Northeastern University
  Ayers, Joe, 27, 110, 111, 113
  Marine Science Center, 110
  Witting, Jan, 110, 113
Nourbakhsh, Illah R. (see CMU)
Nozue family, 224, 225, 226, 227

Office of Naval Research (ONR), 27, 111
OmniMate (see University of Michigan)
Omron Corporation
  Tama, 227
Oppenheimer, Robert J., 32
Osaka University, 58
  Furusho, Junji, 48
  Sakaguchi, Masamichi, 48
  Strut, 48

P3 (see Honda Motor Company)
Paluska, Dan (see MIT)
Paul, Gregory S., 115
Paulos, Eric (see University of California at Berkeley)
Pister, Kris (see University of California at Berkeley)
Pratt, Jerry (see MIT)
Princeton University
  Holmes, Phil, 95
Pyxis Corporation, 186

Quinn, Roger (see Case Western Reserve University)

Raibert, Marc, 24, 28, 29, 39, 86, 183
Real World Interface, Inc.
  Micro ATRV and ATRV-2, 143
RedZone Robotics, Inc., 138, 141
  Rosie, 141
Reona, 79
Rising Sun, 29
Ristow, Christian, 199
Ritzmann, Roy (see Case Western Reserve University)
RoboCup, 82, 83, 214, 215
  Kitano, Hiroaki, 212, 213, 229
Robot Wars, 200, 201, 202, 204, 205
  Mouser Catbot 2000, 205
  Pretty Hate Machine, 200
Robotics and Automation Society of the Institute of Electrical and Electronics Engineers, 46
Robotics Engineering Consortium
  Mini-Dora, 145
  Omniclops, 145
  Schempf, Hagen, 145

Sakai, Masami, 79
Sally Corporation
  Budnik, Lyudamila, 208
  Jacobson, Lance, 210
  Terra M'tica, 208, 210
SARCOS Inc., 21, 50, 52, 54
  Ledger, Charles, 219
  Jacobsen, Steve, 54, 117, 216, 217
  Prue, Doren, 219
  Reynolds, Scott, 219
  SARCOS robot, 21, 218, 219
  Treadport, 137
  Utah Artificial Arm, 216, 217
Scassellati, Brian (see MIT)
Schaal, Stefan (see ATR)
Schempf, Hagen (see Robotics Engineering Consortium)
Schwarzenegger, Arnold, 228
Science University of Tokyo, 74
  Hara, Fumio, 73, 74, 76, 78, 79
  Akasawa, Hidetoshi, 73, 76
  Ayai, Harumi, 78
  Ida, Fumio, 79
  Kobayashi, Hiroshi, 73, 76

Tabata, Masaoki, 73
sea bream, 107
Showa University, 172
Smithsonian magazine, 69
Sony Corporation, 153
  AIBO, 24, 39, 79, 83, 152, 188, 189, 215, 224, 225, 226, 227
Stanford Research Institute, 175
Stanford University, 29, 30, 104, 137
  Bailey, Sean, 98, 99
  Cham, Jorge, 98, 99
  Clark, Jonathan, 99
  Cutkosky, Mark, 98, 99, 137
  Krummel, Thomas, 176
  Mini-Sprawl, 98
  Sprawl, 98
  Sprawlita, 98
Star Wars (defense system), 114
Star Wars, 114, 120, 124, 200
Starlab, 28
Stephenson, Neil
  The Diamond Age, 28
Sun Microsystems, 29, 228

Takanishi, Atsuo (see Waseda University)
Takanobu, Hideaki (see Waseda University)
Tales from the Crypt, 207
Terada, Yuuzi (see Mitsubishi)
Terminator, 37, 228
Terminator 2, 200
Tetsuwan Atomu, 82, 196
Tezuka, Osumu, 196
Three Mile Island, 138, 141
Tilden, Mark, 24, 86, 117, 118, 120, 121
  CTD 1.4, 117
  GPIM 2.4, 117
  Lampbot 1.0, 5
  Nito 1.0, 121
  Spyder, 118
  Unibug 1.0, 2, 240
  Unibug 3.1, 120
  Unibug 3.2, 117
  VBUG 3.3, 117
Tokyo Instit. of Technology, 86, 114
  Blue Dragon, 149
  Hirose, Shigeo, 27, 39, 44, 86, 89, 114, 149, 192
  Hirose-Yoneda Laboratory, 192
  Masahiro, Mori, 196
  Snake-bot ACM R-1, 89
  Titan VII, 192
  Tsukagoshi, Hideyuki, 192
  Vuton, 89
Tsukagoshi Hideyuki (see Tokyo Institute of Technology)
Tsukuba
  Cheng, Gordon, 56, 57
  Electrotechnical Laboratory, 57
  Jack, 56, 57
  Kuniyoshi, Yasuo, 56, 57
  Mechanical Engineering Laboratory, 57
  Shibata, Takanori, 227
Tsukuba Science Exposition of 1985, 37
Tufts University
  Dennett, Daniel C., 25
Turing, Alan, 23
  Turing Test, 24

Unimate, 164
Universal Studios, 217
University of California at Berkeley, 26, 86, 94, 98, 101, 103, 168
  Full, Robert J., 86, 90, 91, 94, 95, 96, 98, 101, 103
  Paulos, Eric, 168
  Poly-PEDAL Laboratory, 90, 91, 96
  Pister, Kris, 25, 26, 27, 32
  PRoP (Personal Roving Presence), 168
  Searle, John, 25, 29
University of Michigan, 95, 96,
  Borenstein, Johann, 189
  Buehler, Martin, 96
  Koditschek, Dan, 95, 96
  LabMate, 189
  OmniMate, 189
  Saranli, Uluc, 96
University of Nagoya, 47
  Brachiator III, 86
  Center for Cooperative Research in Advanced Science and Technology, 46
  Fukuda, Toshio, 46, 47, 86
  Takahashi, Kazuo, 46
University of Reading,
  Warwick, Kevin, 29, 30, 31, 32, 228
University of South Florida, 155
  Murphy, Robin R., 155
University of Southern California, 50, 55
University of Tokyo, 23
  Inoue, Hirochika, 23, 27
University of Utah, 137, 179, 216, 217
  Hollerbach, John M., 137
University of Zurich, 79

van der Waal force, 93
Veloso, Manuela (see Carnegie Mellon University,)
Villa, Alvaro, 207, 208
Vukobratovich, Miomir, 38

Walt Disney Company, 207, 216
Warwick, Kevin (see University of Reading)
  March of the Machines, 223
Waseda University, 14, 18, 39, 43, 172
  Setiawan, Samuel, 38
  Takanishi, Atsuo, 18, 37, 39, 40, 172
  Takanobu, Hideaki, 172
  Universal Dental Robot, 172
  WABIAN-RII, 14, 36, 37, 38
  Wabot 1, 37
  Wabot 2, 37
  WE-3RIII, 40, 233
Wayo Women's University, 172
Westinghouse, 188
Westworld, 200
Whittaker, William L. "Red," (see CMU)
Williams, Ed (see iRobot)
Wired magazine, 29, 30
Wizard of Oz, 117
Wright, Robert
  NonZero: The Logic of Human Destiny, 30

Yamanashi Medical University, 172

Acknowledgments

We thank *Stem* magazine, Germany, for its support and the seed idea for this project. Special thanks to *Stem* technology editor and flying ace Thomas Borchert, our friend and colleague; and picture editor Volker Lensch. In addition, we thank Bob Prior, Michael Sims, Terry Lamoureux, Christine Dunn, and everyone at The MIT Press. Another big thank you to Mike Hawley of the Media Lab for being a grand facilitator.

In addition we thank our staff members Alex Wright, Sheila Foraker, and David Elkins for keeping the office on course during our long absences and for their work during the interminable coffee-drenched hours of book production. As well, we thank our friend and colleague Charles C. Mann, who is slowly learning that the word "editor" is not an epithet, no matter what Dick Teresi says; and designer par excellence David Griffin, who before working with us had never met a chunk of white space he didn't want to preserve and who finally understands that white space is better suited to mountain tops than to book pages. We thank them for their intelligent professionalism, for finding humor when there seemed to be none, and for their incredible lack of memory about the rough spots, which keeps them coming back, book after book.

Thanks to Chelsea Owens for invaluable assistance in transcribing when everyone else was sick to death of it; Katherine Wright for proof-reading and copyediting; and Susan D'Aluisio for the cold-read.

Special thanks to all the scientists, engineers, researchers, research assistants, technicians, administrative assistants, graduate students, and undergrads who allowed us to bother them for an hour that often turned into a day or two.

Thanks also to our family and extended family for their forbearance and encouragement: Josh D'Aluisio-Guerrieri; Jack Menzel; Adam Guerrieri; Evan Menzel; Ray Kinoshita; Newell Mann; Emelia Mann; Kathy Moran; Kyle Griffin.

Special thanks to Sam Hoffman and everyone at San Francisco's most professional and friendly photography lab, The New Lab; as well as Toyoo Ohta and Sanae Lina Wang of Uniphoto Press International, Tokyo; Michael Martin, Rose Taylor, Seymour Yang, and Cordelia Molloy of Science Photo Library, London.

Electric *dreams*: Hans Moravec; Hirochika Inoue; Hugo de Garis; Kevin Warwick; Iain Goodhew; Ben Hutt; Kristofer Pister; Marc Raibert; Marvin Minsky; Rodney Brooks; Annika Pfluger; SARCOS; Takeo Kanade.

Robo *sapiens*: Atsuo Takanishi; Naoki Hieda; Takemitsu Mori; Yohel Nagasaka; Samuel Agus Setiawan; Noriko Okamoto; Masato Hirose; Yasuhisa Hasegawa; Kazuo Takahashi; Toshio Fukuda; Hiroo Mizoguchi; Yoshikuni Ito; Junji Furusho; Masamichi Sakaguchi; Hiroki Kamei; Naoyuki Takesue; Yuuki Kiyota; Takashi Kodera; Stefan Schaal; Shin'ya Kotosaka; Tomohiro Shibata; Mitsuo Kawato; Yasuo Kuniyoshi; Gordon Cheng; Brian Scassellati; Cynthia Breazeal; Paul Fitzgerald; Lijin Aryananda; Mike Binnard; George Homsy; Fumio Hara; Hidetoshi Akasawa; Masaoki Tabata; Harumi Ayai; Norio Kodaira; Hiroshi Kobayashi; Harumi Ayai; Masami Sakai; Tatsuya Matsui; Hiroaki Kitano; Ron Diftler; Tetsuya Ogata; Jerry Pratt; Austin Richards; Shigeki Sugano; Yoji Yamada; Mieko Namba; Honda; Kylie Clark.

Bio *logical*: Shigeo Hirose; Toshio Takayama; Takahisa Yamamoto; Shitomi Yanagawa; Alan DiPietro; Robert J. Full; Martin Buehler; Daniel E. Koditschek; Uluç Saranli; Jorge Cham; Mark Cutkosky; Jonathan Clark; Sean Bailey; John Aspinall; Ed Williams; Roger D. Quinn; Matthew Birch; Richard Bachmann; Gabe Nelson; Roy Ritzmann; James Watson; Luis Pinero; Marc Millis; Yuuzi Terada; John Muir Kumph; Joseph Ayers; Jan Witting; Frank Kirchner; Bernard Klaaflen; Rainer Worst; Giuseppina Campagiorni; Peter Dilworth; Mark Tilden; Jerome Lettvin; Allison Ogden; John Feddema; Brosl Hasslacher; Robert J. White; Randal Beer.

Remote *possibilities*: Michael S. Garrett; Hrand Aghazarian; Terrance L. Huntsberger; Arthur D. Thompson; Robert J. Ambrose; Fredrik L. Rehnmark;Hal A. Aldridge; Bill Bluethmann; Max Fischer; Weston Griffin; Michael Turner; Ryan Findley; Peter Berkelman; John M. Hollerbach; William 'Red' Whittaker; Ben Shamah; James P. Teza; Gary Dimmick; William G. Cogger; Hagen Schempf; Shree Nayar; Drew Bennett; Karen Hawkes; Graham Hawkes; Ronald C. Arkin; Robin Murphy; Paul MacCready; Scott Newbern; Marty Spadaro; Paul Trist Jr.; Tom DeMarino; Carlos Miralles; Omead Amidi; Anne Watzman; Yuri Gawdiak; Ralph Hollis; Richard Petras II; Daniel J. Puetz; Red Zone; Gerd Hirzinger; Michael Krapp; Jürgen Vollmer; Ina Köpke; Hermann Streich; Tom Frost; Adam Peltz; Wyatt C. Sadler; William G. Cogger; RedZone Robotics; Warren Whittaker; Michael Catalan; John G. Watson; JPL (Jet Propulsion Laboratory); Andrea Blacklock; Linda Ryan; Laura Woodbury; Sarah Finney.

Work *mates*: Frank Schad; Yoshihiro Fujita; Aston J. Bridgman; Toshihiro Nishizawa; Eric Paulos; Hideaki Takanobu; Sheila Shah; Volkmar Falk; Thomas Krummel; Anthony M. DiGioia III; Bob Goodman; Harold Sears; Hugh Herr; Dan Paluska; Paul Cappa; Joseph Engelberger; Fred 'Body' Barker; Johann Borenstein; Hideyuki Tsukagoshi; Norio Kodaira; Surgical of Mountain View, California; H. Harry Asada; Yoseph Bar-Cohen; Ulrich Gartner; Intuitive Surgical; Thierry Thaure; Hitoshi Maekawa; Ken Salisbury; Joseph S. Spano; Electrolux Vacuum; Gabriela Schulz; John Hutagalung; Alfred Kärcher GmbH & Co.; Lars Ferner; Gernot Schmierer; Michaela Neuner; NEC Corporation; Ian Kane; Sundar Vedula; Peter Rander; Priyan Gunatilake; Huodong Wu; Mel Siegel; Hideo Saito.

Serious *fun*: Christian Ristow; SRL; Christian S. Carlberg; Marc Thorpe; Fon Davis; April Mousley; David Hall; Alvaro Villa; AVG; Juan Valencia; Lyudmila Budnik; Jan Sherman; Sally Corporation; Lance Jacobson; Ben Edwards; Rich Hill; Manuela Veloso; Peter Stone; Michael Bowling; Sorin Achim; Elly Zoe Winner; Stephen C. Jacobsen; Abhijeet Vijayakar; Ryan C. Hayward; Rodney Freier; Marlene Osoro; Andrea Seigberg; Peter Stone; Fraser Smith; Scott Reynolds; Doren Prue; Charles Ledger; James Kolzs; Illah R. Nourbakhsh; Sony CSL; Carolyn E. O'Brien; The Nozue Family; Takanori Shibata; Colin Angle; iRobot; Clayton Bailey; Betty G. Bailey; GMD Robocup (Germany); Jörg Wilberg; Peter Schöll; Herbert Jaeger; Hans-Ulrich Kobialka; Helen Greiner; Daniel Lintz; Robot Wars; Andrew Bennett; Mark Hays; Robosaurus; Richard M. Scott; Hasbro, Inc.

Library of Congress Cataloging-in-Publication Data

Menzel, Peter, 1948– .
    Robo sapiens : evolution of a new species / Peter Menzel and Faith D'Aluisio.
        p. cm.
    "A Material world book."
    Includes bibliographical references and index.
    ISBN 0-262-13382-2 (hc. : alk. paper)
    1. Robotics. 2. Artificial intelligence. 3. Intelligent control systems.
I. D'Aluisio, Faith, 1957– . II. Title.
TJ211 .M45 2000
629.8'92—dc21   00-033946

Prairie sunflower, *Helianthus petiolaris*,

designed by nature.

Unibug 1.0, designed by Mark Tilden.